Analog, Digital and Multimedia
Telecommunications

Analog, Digital and Multimedia Telecommunications

BASIC AND CLASSIC PRINCIPLES

Omar Fakih Hamad

Xlibris Corporation
Victory Way, Admirals Park Crossways
Dartford DA2 6QD, United Kingdom

FIRST EDITION
for university and college students

ISBN: Softcover   978-1-4568-1020-7
               Ebook   978-1-4568-1021-4

Hamad, Omar Fakih, 1971-
Analog, Digital and Multimedia Telecommunications:
Basic and Classic Principles

This book was printed in the United States of America.

To order additional copies of this book, contact:
Xlibris Corporation
0-800-644-6988
www.xlibrispublishing.co.uk
Orders@xlibrispublishing.co.uk
301065

Contents

Dedicated to:

His Excellency Juma Hamad Omar, former principal of Fidel Castro Secondary School in Zanzibar and former minister in the United Republic of Tanzania, for his endless efforts in promoting higher education to intelligent, hardworking, poor students who could, otherwise, not be able to make it financially!

Preface

In spite the fact that the presentation of this elementary reference book has fetched abundant knowledge and ideas from different sources and people, this book, by itself, is a self referencing piece of work for the beginners and intermediate learners in the field of telecommunications engineering and sciences.

The book has been the result of experiences gained from teaching and learning the materials of the courses—individually and collectively—as analog telecommunications, digital telecommunications and multimedia telecommunications at undergraduate and postgraduate levels.

I am very grateful to the students and the staff for their enormous contributions, motivation and encouragement to make the preparation of this book possible. Many conversations and collaborations with different staff, students and friends in the industry and teaching and learning arenas have been the key success to the book—the success that it owes them big. The direct and indirect encouragements from the head and founder of the telecommunications engineering at the University of Dar es Salaam, Dr. M. M. Kissaka, and the rest of telecommunications engineering staff of putting the materials into a book have been the fundamental inspiration.

The third part of the book, multimedia telecommunications, has been much successful because of the several fruitful discussions that I have had with my friends—seniors and juniors—in Multimedia Data Communications Laboratory and specifically very close discussions with my PhD professor and supervisor, Dr. Ji Seung Nam, of the School of Electronics and Computer Engineering at Chonnam National University in the Republic of South Korea.

The real preparation of the manuscript started long back in July 2002 when I started teaching the course to our students at the University of Dar es Salaam, and the process continued till in 2010 when the University of Johannesburg, through the Executive Dean of the Faculty of Engineering and the Built Environment, Professor Tshilidzi Marwala, empowered the process through a post doctoral research fellowship and grant offered to me—in collaboration between the University of Johannesburg and the South African National Research Foundation (NRF).

This first edition of the book—for university and college students—has been purposely divided into three parts. The basic and classic principles of analog, digital and multimedia telecommunications have been addressed in a simplified manner and without a need of cross-referencing of the encountered equations. That has been done for students to get the direct and inter-related flow of the materials as they read and practice the examples included and those which have been borrowed from other literatures. In addition to the three appendices on tutorials, quizzes and tests, the book has been further divided into twenty three short and manageable chapters including the chapter on conclusions, discussions and future direction on multimedia telecommunications, particularly based on multimedia data delivery challenges.

Part I—Analog Telecommunications—is comprised of nine chapters with Chapter One giving Introduction to Telecommunications; Chapter Two illustrating the need for Signals and Spectra in Telecommunications; Chapter Three on Liner Continuous Wave (CW) Modulation; and Chapter Four devoted to Exponential Continuous Wave (CW) Modulation. Chapter Five is on Noise in Telecommunications Systems; Chapter Six explains Multiplexing; Chapter Seven presents ides on Analog Pulse Modulation; Chapter Eight is about analog based Pulse-Code Modulation; and Chapter Nine touched ideas on Data Transmission.

Part II—Digital Telecommunications—has eight chapters where Chapter Ten gives Introduction to Digital Telecommunications; Chapter Eleven is about digital based Pulse Code Modulation (PCM); Chapter Twelve present principles of Delta Modulation; Chapter Thirteen talks about Phase Shift Keying (PSK); Chapter Fourteen is on Frequency Shift keying (FSK); Chapter Fifteen discusses about Quadrature Amplitude Modulation (QAM); Chapter Sixteen is about Digital Multiplexing Techniques and Hierarchies; and Chapter Seventeen gives detailed discussion on techniques, theories and principles on Information Theory and Coding.

Part III—Multimedia Telecommunications—has six chapters with Chapter Eighteen giving Introduction to Multimedia Telecommunications; Chapter Nineteen discussing the Approach to

Multimedia Delivery; Chapter Twenty discussing the Bandwidth-Latency-Product Measures (BLPMs); Chapter Twenty One describing the NGS-Based Max-Heap Overlay Multicast Scheme; Chapter Twenty Two illustrating the Performance Evaluation of Max-Heap Overlay Tree; and Chapter Twenty Three explaining the Conclusions, Discussion and Possible Future Direction in multimedia delivery techniques.

Omar Fakih Hamad
University of Johannesburg
South Africa

Part I
Analog Telecommunications

Chapter One

Introduction to Telecommunications

1.1 General introduction

Telecommunications, implying electronic communication, defines the science and technology of transmitting information electronically by wires, cables, fibres or radio signals with integrated encoding and decoding equipment. It is about information transmission and switching over communications lines or channels. Technically, information transmission and switching are what make information transfer. It is to be made clear that, at a broader translation, information transfer incorporates both, information transmission and device switching. Nevertheless, in this elementary and introductory edition of the subject, the two will be, and can be, used interchangeably and sometimes complementing each other.

With that basic definition in mind, the telecommunications engineer's main concern is confined within transmission reception of information signals. By a signal, we mean an electric voltage or current which varies with time and is used to carry messages or information from one point to another. The form of a message in telecommunications engineering can be either in word or coded symbols. However, of great importance in telecommunications is the amount of information the message contains.

But why are we interested in converting information into a signal? It is mainly because it is more convenient to handle information as signal. Through transmission channels, the signal is easily transmitted over a communication system where, at the destination, the signal can then be transformed back to the original information or message.

1.2 What is telecommunications?

In basic electric sense, telecommunications is the sending, receiving, and processing of information by electric means. The evolution of telecommunications and its use emerged as early as in the dates before the nineteenth century, but the documentation started in 1840s with wire telegraphy coming into use. The invention of telephony was observed in the 1870s while in 1900s the field had enjoyed the birth of radio technologies with triode tube which made a great use in WWII. The success gained during the WWII made the field to be widely used further and more refined through the invention of the transistor and the use of the transistor, IC, and other semiconductor devices.

More recently, the use of satellites and fibre optics has made telecommunications more widespread with increasing emphasis on computer and other data communications devices. A modern telecommunications system is concerned with the sorting, the processing, the storing, transmission, further processing, the filtering of noise and the reception with the processing steps including decoding, storage, and interpretation of the signals into real world data.

The forms of telecommunications include radio telephony and telegraphy, broadcasting, point-to-point, mobile communications, computer communications, radar systems, radio-telemetry and radio aids to navigation, and so many others emerging technologies like overlay multicasting and peer-to-peer multimedia sharing over the internet.

Telecommunications background empathizes on proper and adequate understanding of the operations of amplifiers and oscillators, the building blocks of them, and the electronic processes and equipment involved. It is also important to understand the daily telecommunications concepts like noise, modulation, and information theory. Logically, it is a wise and suitable idea to considered basic systems, communications process and circuits, and more complex systems and in that order.

1.3 Very basic telecommunications system

Considering the fact that telecommunications is about sending, processing and receiving of information from a source to at least one destination, Fig. 1.1 depicts a very basic telecommunications system an information source and a destination exchanging information via a communication channel.

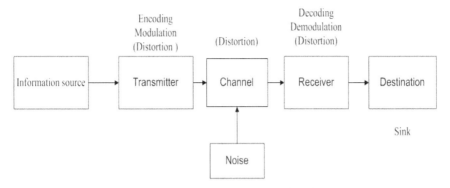

Fig. 1.1. A very basic telecommunications system

At the source, the message signals are generated for the transmitter where the transmitter process the signals and send them along the transmission line or channel to the receiver. The receiver extracts the messages and sends tem to their final destination or sink. During the process, the noise is picked up from various sources during transmission and reception.

1.4 A telecommunications system with input and output transducers

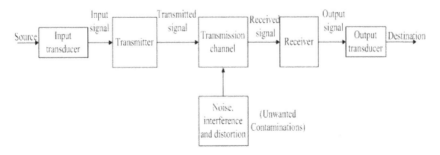

Fig. 1.2. Elements of a telecommunication system in a voice communication system

Fig. 1.2 describes a typical telecommunications system in a voice communication system where we need to have an input transducer which can be thought of as7, and in fact it is, a microphone, while the output transducer can be a conventional loud speaker

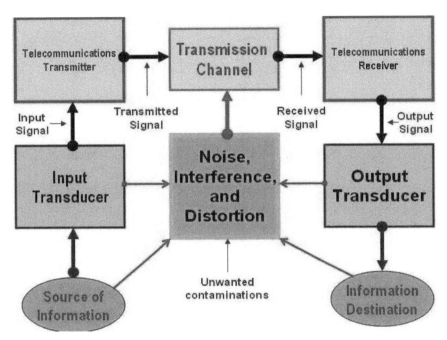

Fig. 1.3. A telecommunication system
with input and output transducers

Fig. 1.3 shows a more generalized telecommunications system with all the basic elements and the external noise environment shown with its effects towards the whole system. It is shown that source of information itself might be a contributor towards the system's overall nose. The source submits the real world like information to an input transducer whereby the transducer itself might also have some effects to noise contribution. The telecommunications transmitter accepts the signals to transmission channel which, along with the noise from the transmitter, it has its own sources of noise affecting the transmission. The received signals from the channel are accepted by the telecommunications receiver before transforming the signals into output transducer compatible format ready to be sent to the information destination which might also have some contributions towards the overall performance of the received information.

Three essential parts of any telecommunications system are unavoidable. These are the transmitter, the transmission channel, and the receiver. The transmitter processes the input signal to produce a transmitted signal suited to the characteristics of the transmission channel. Therefore, signal transmission involves modulation and coding. The transmission channel is basically the electrical or rather electromagnetic medium that bridges the distance from source to destination, that is, transmitter to receiver. This may be a pair of wires, a coaxial cable, a radio wave, laser beam, or an optical fibre. Every channel experiences some amount of transmission loss which we refer as attenuation. In addition, it is known that signal power is inversely proportional to the transmission distance, that is, it can expressed as in Eq. (1.1): -

$$\text{Signal Power} = \alpha \frac{1}{dis\tan ce} \quad \text{--(1.1)}$$

Therefore, the receiver is dealing with the preparation for delivery to the destination. The receiver operations include amplification, demodulation, decoding, and filtering. Nevertheless, telecommunications systems will always experience attenuation, noise distortion, and interference. While attenuation reduces signal strength at the receiver, on one side, noise, distortion, and interference, on the other hand, lead to alteration of the signal shape. Even though, there might be several known sources of this, the convention is to blame them entirely on the channel.

1.5 Information, messages, and signals

Information in telecommunications signifies semantic and philosophical notions that defy precise definition while message implies physical manifestation of information as produced by the source. Signals are the detectable electrical pulses representing messages or information. They are in voltage or current form. Signal and message are sometimes used interchangeably. They are both physical embodiment of information.

Chapter Two

Signals and Spectra in Telecommunications

2.1 Introduction

Signals, meant in telecommunications, are *time-varying* quantities such as voltages and currents, which are used to carry information from a point to another. Although signals physically exist in the *time domain,* they can also be represented in *frequency domain.* This means that, signals can be viewed as consisting of *sinusoidal* components at *various frequencies.*

Spectrum is in short the frequency domain description of the signal. Spectral analysis is customarily done by Fourier series and Fourier Transform. The analysis makes the fundamental method of telecommunications engineering. It allows one to treat entire classes of signals that have similar properties in the frequency domain avoiding detailed time domain analysis of individual signals. The spectral analysis approach provides valuable insight for design work in telecommunications systems. Therefore, it is advised that a special attention to the frequency domain interpretation of signal properties must be given.

2.2 Types of signals in telecommunications

There are two main types of signals—the analogue signals which vary continuously with time and the digital signals which are discontinuous with time.

2.2.1 Analogue signals

Usually, analog signals represent the variation of a physical quantity. Example includes a sound wave. They are either single sine waves or a combination of them. A simple analogue signal is shown in Fig. 2.1 where the voltage quantity *V* varies continuously with time *t*.

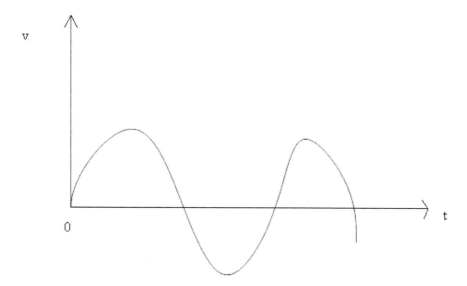

Fig.2.1. A simple analogue signal

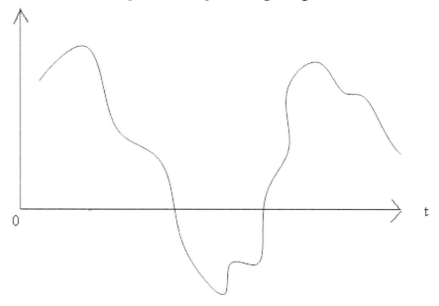

Fig.2.2. Analogue signal

An example depicted in Fig.2.1 shows certain nature of uniformity and predictability, but in Fig.2.2 the example of analogue signal is shown where by the quantity on the vertical line, be it current or voltage against time *t*, can not predicted with a simple guess. Both, Fig.2.1 and Fig.2.2, represents analogues signals in their very general term.

2.2.2 Digital signals

These consist of pulses occurring at discrete intervals of time. The pulses may occur singly with a definite periodicity or they may occur in group in the form of code.

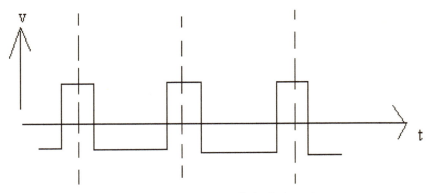

Fig.2.3. A simple digital signal

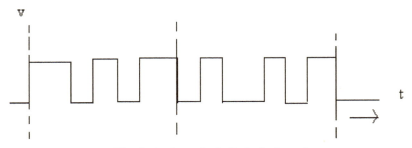

Fig.2.4. A coded digital signal

Fig. 2.3 shows pulses that occur singly with a definite periodicity while the signals in Fig. 2.4 correspond to signals occurring in groups in the form of code.

Examples of signals commonly used in our daily telecommunications applications include telegraphy, telephony, radio communication, television, and radar signals.

2.3 Spectrum of a signal

We prefer to have our signals into possible frequency components. Signals can be analyzed by Fourier techniques into various frequency components. The frequency spectrum of the signal reflects the total range of the components' frequencies. This is of prime importance in telecommunications systems and engineering.

An exact knowledge of such a spectrum is useful in solving problems of transmission and reception in telecommunications engineering and design.

2.3.1 Signal representation

In the time domain a signal is represented with a plot of instantaneous amplitude against time while in the frequency domain, it is represented with a plot of its spectral component amplitudes against frequency. There is a direct link between these two representations and this is obtained with the aid of Fourier transforms. There are, in principle, two kinds of spectra—the discrete spectra and the continuous spectra.

If we consider periodic signals, v, as the signals that obey the relationship expressed in Eq. (2.1) as: -

$$v(t \pm mT_o) = v(t) \text{ ....} (2.1)$$

where T_o = the repetition period = $\frac{2\pi}{\omega_0}$, $-\infty < t < \infty$, and $m \in \mathbb{Z}$. It means that shifting the signal by an integer number of periods to the left or right leaves the waveform unchanged. That leaves a freedom to say that a periodic signal is fully described by specifying its be-havior over any one period. Fig.2.5 describes an example of a peri-odic signal with time period T.

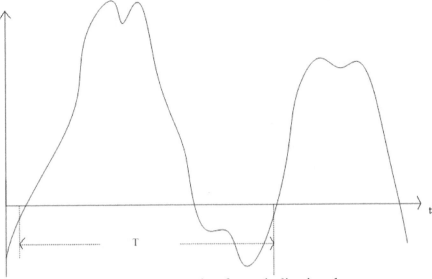

Fig. 2.5. Example of a periodic signal

2.4 Representation of a periodic signal by the Fourier series

Any periodic function of time, $v(t)$, can be represented by the Fourier series expression presented in Eq. (2.2)

$$f(t) = v(t) = a_0 + \sum_1^\infty a_n Cos(n\omega t) + \sum_1^\infty b_n Sin(n\omega t) \dots\dots (2.2)$$

where a_n and b_n are the coefficients to be evaluated such as in Eqs. (2.3) and (2.4):

$$a_n = \frac{2}{T}\int_{-T/2}^{+T/2} v(t)Cos(n\omega t)dt \dots\dots (2.3)$$

$$b_n = \frac{2}{T}\int_{-T/2}^{+T/2} v(t)Sin(n\omega t)dt \dots\dots (2.4)$$

where $\omega = \dfrac{2\pi}{T}$ and T is a periodic time.

The coefficient a_0 corresponds to the *dc* term which is the average of $v(t)$ in a period T. Mathematically, this can be expressed as in Eq. (2.5):

$$a_0 = \frac{1}{T} \int_{T/2}^{T/2} v(t)dt \quad\text{---} \quad (2.5)$$

It must be noted that if $v(t) = v(-t)$, it means that we have an even function and that there is symmetric about origin and only *Cosine* terms will be present with optional *dc* components. Otherwise, if $v(t) = -v(-t)$, it means that we have an odd function and that only *Sine* terms are present with optional *dc* components. In addition, if $v(t + T/2) = v(t)$, it means that we have only even harmonics present while if $v(t + T/2) = -v(t)$, it means that we have only odd harmonics present.

As an **example, let us c**onsider a square-wave signal as in Fig. 2.6 that varies between the values of $+1V$ and $-1V$. -

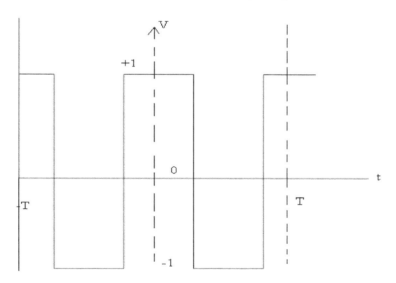

Fig. 2.6. A square-wave signal

If the periodic time is T and is symmetrical with respect to the vertical axis at time *t=0*. To obtain the Fourier components of the waveform can be dealt with starting with the computation of a_n as in Eq. (2.3) since $b_n = 0$. If, in addition, the waveform is also symmetrical about the horizontal axis, it implies that the average area is zero and that the *dc* term, $a_0 = 0$.

It can be verified that the functions $v(t) = v(-t)$ and, hence, only *Cosine* terms are present. This means that $b_n = 0$ as it can clearly be visualized.

Recalling the expression for a_n that $a_n = \dfrac{2}{T} \int_{-T/2}^{+T/2} v(t) Cos(n\omega t) dt$

where, based on the given square-wave signal, v(t) values at different ranges are: -

$$v(t) = -1 \, from \; {}^{-T}\!/\!_2 \; to \; {}^{-T}\!/\!_4$$
$$v(t) = +1 \, from \; {}^{-T}\!/\!_4 \; to \; {}^{+T}\!/\!_4$$
$$v(t) = -1 \, from \; {}^{+T}\!/\!_4 \; to \; {}^{+T}\!/\!_2$$

Therefore, for the three significant ranges, a_n can be expressed as in Eq. (2.6)

$$a_n = \frac{2}{T} \left\{ \int_{-T/2}^{-T/4} -Cos(n\omega t)dt + \int_{-T/4}^{+T/4} Cos(n\omega t)dt - \int_{+T/4}^{+T/2} Cos(n\omega t)dt \right\}$$

--- (2.6)

$$= \frac{2}{T} \left\{ \left[-\frac{Sin(n\omega t)}{n\omega} \right]_{-T/2}^{-T/4} + \left[\frac{Sin(n\omega t)}{n\omega} \right]_{-T/4}^{+T/4} - \left[\frac{Sin(n\omega t)}{n\omega} \right]_{+T/4}^{+T/2} \right\}$$

$$= \frac{2}{n\omega T} \left\{ \begin{array}{l} -Sin\left(\dfrac{-n\omega T}{4}\right) + Sin\left(\dfrac{-n\omega T}{2}\right) + Sin\left(\dfrac{n\omega T}{4}\right) \\ -Sin\left(\dfrac{-n\omega T}{4}\right) - Sin\left(\dfrac{n\omega T}{2}\right) + Sin\left(\dfrac{n\omega T}{4}\right) \end{array} \right\}$$

Therefore,

$$a_n = \frac{8}{n\omega T} Sin\left(\frac{n\omega T}{4}\right) - \frac{4}{n\omega T} Sin\left(\frac{n\omega T}{2}\right) \quad \text{------------------ (2.7)}$$

Replacing $\omega T = 2\pi$ in Eq. (2.7), we get simplified a_n as in Eq. (2.8):

$$a_n = \frac{8}{2n\pi} Sin\left(\frac{n\pi}{2}\right) - \frac{4}{2n\pi} Sin(n\pi) \dots\dots (2.8)$$

However, $Sin(n\pi) = 0$ and, therefore, a_n gets further simplified into Eq. (2.9):

$$a_n = \frac{8}{2n\pi} Sin\left(\frac{n\pi}{2}\right) = \frac{4}{n\pi} Sin\left(\frac{n\pi}{2}\right) \dots\dots (2.9)$$

Hence, we have a_0, a_1, a_2 and a_3 given as in Eqs. (2.10), (2.11), (2.12), (2.13), respectively.

$$a_0 = 0 \ (dc \ term) \dots\dots (2.10)$$

$$a_1 = \frac{4}{\pi} Sin\left(\frac{\pi}{2}\right) = \frac{4}{\pi} \dots\dots (2.11)$$

$$a_2 = \frac{4}{2\pi} Sin(\pi) = 0 \dots\dots (2.12)$$

$$a_3 = \frac{4}{3\pi} Sin\left(\frac{3\pi}{2}\right) = \frac{-4}{3\pi} \dots\dots (2.13)$$

Similarly, the other terms can be found in the same fashion.

Now, from the expression for $f(t)$, we again produce Eq. (2.14) as:

$$f(t) = v(t) = a_0 + a_1 Cos(\omega t) + a_2 Cos(2\omega t) + \dots \dots\dots (2.14)$$

Therefore, substituting the values for a_0, a_1, a_2 and a_3 we get Eq. (2.15)

$$f(t) = v(t) = \frac{4}{\pi}\left(Cos(\omega t) - \frac{1}{3}Cos(3\omega t) + .\frac{1}{5}Cos(5\omega t) - ... \right) \quad (2.15)$$

2.4.1 Typical series in use

The Fourier components of typical waveforms used in practice are shown in Eqs. (2.16) through (2.19) along with Figs. 2.7 through 2.10.

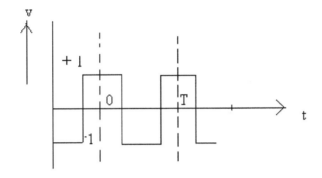

$$f(t) = \frac{4}{\pi}\left(Cos(\omega t) - \frac{1}{3}Cos(3\omega t) + .\frac{1}{5}Cos(5\omega t) - ... \right) \text{............} (2.16)$$

Fig. 2.7. Symmetric square wave

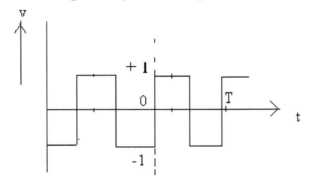

$$f(t) = \frac{4}{\pi}\left(Sin(\omega t) + \frac{1}{3}Sin(3\omega t) + .\frac{1}{5}Sin(5\omega t) + ... \right) \text{............} (2.17)$$

Fig. 2.8. Asymmetric square wave

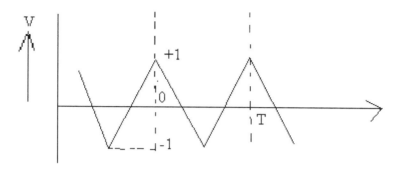

$$f(t) = \frac{8}{\pi^2}\left(Cos(\omega t) + \frac{1}{9}Cos(3\omega t) + .\frac{1}{25}Cos(5\omega t) + ... \right) \text{------} (2.18)$$

Fig. 2.9. Triangular wave

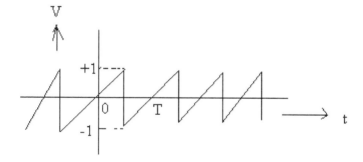

$$f(t) = \frac{2}{\pi}\left(Sin(\omega t) - \frac{1}{2}Sin(2\omega t) + .\frac{1}{3}Sin(3\omega t) - ... \right) \text{------------} (2.19)$$

Fig. 2.10. Saw-tooth wave

The series mentioned along with Figs. 2.7 through 2.10 can be readily used while resolving any signal that corresponds to the respective series. Else, most of the signals can be decomposed into separate functions whose series have here easily kept available.

2.5 Complex representation of signal

An alternative, but convenient way, of writing the periodic functions or signals, let say, *f(t)* is in terms of complex quantities. From complex signal theory, we have the expressions for $Cos(n\omega t)$ and $Sin(n\omega t)$ in terms of real and imaginary terms as in Eqs. (2. 20) and (2. 21):

$$Cos(n\omega t) = \frac{e^{jn\omega t} + e^{-jn\omega t}}{2} \quad\text{---------------------------------------} \quad (2.20)$$

and

$$Sin(n\omega t) = \frac{e^{jn\omega t} - e^{-jn\omega t}}{2j} \quad\text{-----------------------------------} \quad (2.21)$$

From the original expression for the Fourier series of *f(t)*, as repeated in Eq. (2.22)

$$f(t) = a_0 + \sum_1^\infty a_n Cos(n\omega t) + \sum_1^\infty b_n Sin(n\omega t) \quad\text{-----------------} \quad (2.22)$$

Substituting the complex expressions for $Cos(n\omega t)$ and $Sin(n\omega t)$, then we can have Eq. (2.23),

$$f(t) = a_0 + \sum_1^\infty a_n\left(\frac{e^{jn\omega t} + e^{-jn\omega t}}{2}\right) + \sum_1^\infty b_n\left(\frac{e^{jn\omega t} - e^{-jn\omega t}}{2j}\right) \quad\text{------}(2.23)$$

$$= a_0 + \sum_1^\infty \left\{ \frac{(a_n - jb_n)e^{jn\omega t}}{2} + \frac{(a_n + jb_n)e^{-jn\omega t}}{2} \right\}$$

Putting: -

$$C_n = \frac{(a_n - jb_n)}{2}; \quad C_{-n} = \frac{(a_n + jb_n)}{2}; \text{ and } C_0 = a_0,$$ where C_{-n} is the complex conjugate of C_n

From:

$$a_n = \frac{2}{T} \int_{-T/2}^{+T/2} f(t)Cos(n\omega t)dt$$

and

$$b_n = \frac{2}{T} \int_{-T/2}^{+T/2} f(t)Sin(n\omega t)dt$$

Substituting, we get Eqs. (2.24) and (2.25) or Eqs. (2.26) and (2.27).

$$C_n = \frac{1}{T} \int_{-T/2}^{+T/2} f(t)[Cos(n\omega t) - jSin(n\omega t)]dt \quad\text{(2.24)}$$

and

$$C_{-n} = \frac{1}{T} \int_{-T/2}^{+T/2} f(t)[Cos(n\omega t) + jSin(n\omega t)]dt \quad\text{(2.25)}$$

or

$$C_n = \frac{1}{T} \int_{-T/2}^{+T/2} f(t)e^{-jn\omega t}dt \quad\text{(2.26)}$$

and

$$C_{-n} = \frac{1}{T} \int_{-T/2}^{+T/2} f(t)e^{jn\omega t}dt \quad\text{(2.27)}$$

This implies that $f(t)$ can now be expressed as in Eq. (2.28) or Eq. (2.29).

$$f(t) = C_0 + \sum_1^{\infty} C_n e^{jn\omega t} + \sum_{-\infty}^{-1} C_n e^{jn\omega t} \quad\text{(2.28)}$$

Therefore,

$$f(t) = \sum_{-\infty}^{+\infty} C_n e^{jn\omega t}$$ --- (2.29)

The observation above implies that the periodic signal or function, $f(t)$, can also be represented mathematically by an infinite set of positive and negative frequency components. The negative frequencies have a mathematical significance and they may sometimes also have physical significance. As a positive frequency, it can be associated with an anticlockwise rotation and a negative frequency with a clockwise rotation.

As an example, let us deduce the Fourier series which corresponds to the waveform of a positive-going rectangular pulse train, each pulse of duration τ, repetitive in the period T, and with amplitude E as depicted in Fig. 2. 11.

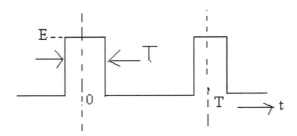

Fig. 2. 11. Positive going rectangular pulse train

We solve this by presuming the general Fourier series for the pulse train given in Fig. 2.11 as given by the general *f(t)* expression as in Eq.: -

$$f(t) = a_0 + \sum_{1}^{\infty} a_n Cos(n\omega t) + \sum_{1}^{\infty} b_n Sin(n\omega t)$$ -------------------- (2.30)

where, after substituting the proper levels and given parameters, we get Eq. (2.31)

$$a_0 = \frac{1}{T} \int_{-T/2}^{+T/2} f(t)dt = \frac{1}{T} \int_{-\tau/2}^{+\tau/2} E dt = \frac{E}{T} [t]_{-\tau/2}^{+\tau/2} = \frac{E\tau}{T} \quad \text{-----------------} \quad (2.31)$$

$$b_n = 0$$

and

$$a_n = \frac{4E}{n\omega T} Sin\left(\frac{n\omega T}{2}\right)$$

Therefore, we have the function expressed by Eq. (2.32).

$$f(t) = \frac{E\tau}{T} + 2\frac{E\tau}{T} \sum_{1}^{\infty} \frac{Sin(n\omega\tau/2)}{(n\omega\tau/2)} Cos(n\omega t) \quad \text{-------------------} \quad (2.32)$$

Dealing with Fourier transform for continuous spectra will have a special importance in understanding the signal characteristics in telecommunications systems. We know that periodic signals last forever, theoretically while non-periodic signals are concentrated over relatively short-time duration. If a non-periodic signal has finite total energy, its frequency-domain representation will be a continuous spectrum obtained from the Fourier transform. This is discussion of the next section.

2.6 Fourier Transforms

Consider the two typical non-periodic signals given in Figs. 2. 12 and 2. 13 where Fig. 2.12 represents a strictly time-limited non-periodic signal and Fig. 2.13 represents an asymptotically time limited signal.

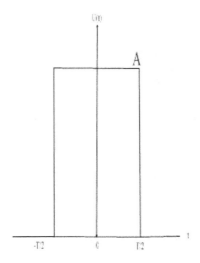

$v(t) = 0$ outside the pulse duration.

Fig. 2. 12. Strictly time-limited non-periodic signal

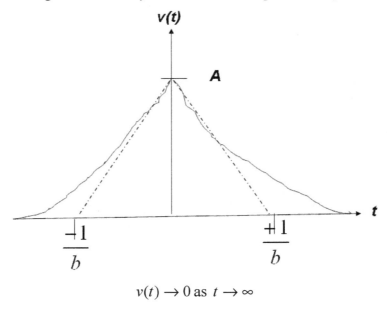

$v(t) \to 0$ as $t \to \infty$

Fig. 2.13. Asymptotically time limited signal

Such signals may also be described loosely as *"pulses"*. The average of $v(t)$ or $|v(t)|^2$, over all time approaches zero. Therefore, a more meaningful property of a non periodic signal is its energy. The normalized signal energy E is defined by Eq. (2.33) as: -

$$E \triangleq \int_{-\infty}^{+\infty} |v(t)|^2 dt \qquad\qquad (2.33)$$

Example, for a rectangular pulse with amplitude A, the energy is $E = A^2 \tau$. This can easily be computed by following the simple steps just described.

When $\int_{-\infty}^{+\infty} |v(t)|^2 dt$ exists and yields energy, E, such that $0 < E < \infty$, then the signal $v(t)$ is said to have well-defined energy and will be called non-periodic energy signal. Almost all time-limited signals of practical interest fall under this category, which is the essential condition for spectral analysis using the Fourier transform.

Starting with the Fourier series representation of a periodic power signal given in Eq. (2.34): -

$$v(t) = \sum_{n=-\infty}^{\infty} C(nf_0) e^{j2\pi nf_0 t} \qquad\qquad (2.34)$$

$$= \sum_{n=-\infty}^{\infty} \left[\frac{1}{T_0} \int_{T_0} v(t) e^{-j2\pi nf_0 t} dt \right] e^{j2\pi nf_0 t}$$

We can note that there is a similar representation for a non periodic energy signal. For a non-periodic energy signal, we have Eq. (2.35),

$$v(t) = \int_{-\infty}^{+\infty} \left[\int_{-\infty}^{+\infty} v(t) e^{-j2\pi ft} dt \right] e^{j2\pi ft} df \qquad\qquad (2.35)$$

The inner part of integration is what is called the Fourier transform of $v(t)$ denoted by $V(f)$ or $£[v(t)]$ and expressed in Eq. (2.36) as:

$$V(f) = \pounds[v(t)] \underline{\underline{\Delta}} \int_{-\infty}^{+\infty} v(t)e^{-j2\pi ft}\, dt \quad\text{(2.36)}$$

The operation in Eq. (2.36) yields a function of the continuous variable f. It is customarily to denote the Fourier transform pair $(v(t), V(f))$ as in Eq. (2.37)

$$(v(t) \leftrightarrow V(f)) \quad\text{(2.37)}$$

where the forward direction implies the Fourier transform of $v(t)$ to $V(f)$, and the backward direction implies the inverse Fourier transform of $V(f)$ to $v(t)$.

We can, therefore, denote the inverse Fourier transform of $V(f)$ to $v(t)$ by Eq. (2.38) as:

$$v(t) = \pounds^{-1}[V(f)] \underline{\underline{\Delta}} \int_{-\infty}^{+\infty} V(f)e^{j2\pi ft}\, df \quad\text{(2.38)}$$

Eq. (2.38) reflects integration over all frequency range, f.

We, hence, have a constitution of the pair of Fourier integrals made of Eqs. (2.36) and (2.38) and that if $V(f)$ is known, then $v(t)$ can be found from (2.38). Similarly, if $v(t)$ is known, then $V(f)$ can be found from (2.36). $V(f)$ is the spectrum of the non-periodic signal $v(t)$. Therefore, a non-periodic signal will have a continuous spectrum, rather than a line spectrum.

$V(f)$ has three major properties worth to not and appreciate. These are:

(i)   the Fourier transform is a complex function. So, $|V(f)|$ is the amplitude spectrum of $v(t)$ and $\arg V(f)$ is the phase spectrum;

(ii)   the value of $V(f)$ at $f = 0$ equals the net area of $v(t)$, since $V(0) = \int_{-\infty}^{+\infty} v(t)dt$; and

(iii)   if $v(t)$ is real, then $V(-f) = V^*(f)$ and $|V(-f)| = |V(f)|$. This means even amplitude symmetry. Similarly, $\arg V(-f) = -\arg V(f)$. This means odd phase symmetry.

The two conditions give a symmetry commonly known as *hermitian symmetry.*

Let us take an example of a rectangular pulse given in Fig. 2.14, and let us adopt the pictorial notation such that the rectangular pulse be represented as in Eq. (2.39):

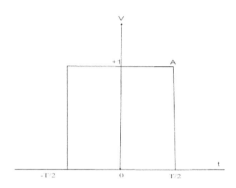

Fig. 2.14. A rectangular pulse

$$\Pi(t/\tau) \underset{=}{\triangle} \begin{cases} 1, |t| < \frac{\tau}{2} \\ 0, |t| > \frac{\tau}{2} \end{cases}$$ (2.39)

The representation means a rectangular function with unit amplitude and duration τ centred at $\tau = 0$. It represents a specific case of a function $v(t) = A\Pi(t/\tau)$

Recalling Eqs. (2.36) and (2.38) we re-write Eq. (2.40) as:

$$V(f) = \pounds[v(t)] \underset{=}{\triangle} \int_{-\infty}^{+\infty} v(t)e^{-j2\pi ft} dt$$ (2.40)

For our rectangular pulse, this means that:

$$V(f) = \int_{-\frac{\tau}{2}}^{+\frac{\tau}{2}} Ae^{-j2\pi ft} dt = \frac{A}{\pi f} Sin(\pi f\tau) = \frac{A\tau}{\pi f\tau} Sin(\pi f\tau) = A\tau Sinc(f\tau)$$ (2.41)

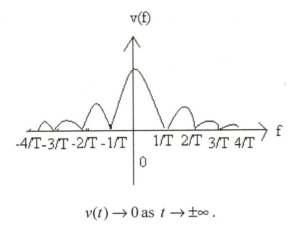

$$v(t) \rightarrow 0 \text{ as } t \rightarrow \pm\infty.$$

Fig. 2.15.Asymptotically time-limited non-periodic signal

The factor $1/\tau$ in Eq. (2.41) is a measure of the spectral width. This depicts the reciprocal spreading of the spectral which indicates that long pulses have narrow spectra and vice versa!

2.7 Fourier transform in terms of ω

Extending the Fourier series technique to non-periodic signals by making $T \rightarrow \infty$. Adjacent pulses virtually never occur. This implies that the pulse train becomes just as a single isolated pulse. Let $f(t)$ be initially periodic. We, therefore, have Eq. (2.42) given as

$$f(t) = \sum_{-\infty}^{+\infty} C_n e^{jn\omega t} \qquad (2.42)$$

where

$$C_n = \frac{1}{T} \int_{-T/2}^{+T/2} f(t)e^{-jn\omega t} dt \qquad (2.43)$$

For a single pulse, we have $T \rightarrow \infty$ and, hence, with $\omega = \dfrac{2\pi}{T}$, we can find that ω tends to be a small quantity $d\omega$. From the Eq. (2.42) for ω, we can have Eq. (2.44)

$$\frac{1}{T} = \frac{\omega}{2\pi} \rightarrow \frac{d\omega}{2\pi} \quad \text{..} \quad (2.44)$$

For the same reasoning, we have Eq. (2.45) given as:

$$n\omega \rightarrow nd\omega \rightarrow \omega \quad \text{..} \quad (2.45)$$

Mathematically, in the limit, the sigma sign leads to an integral sign. Therefore, we can have Eqs. (2.46) and (2.47) as:

$$C_n = \frac{d\omega}{2\pi} \int_{-\infty}^{+\infty} f(t)e^{-j\omega t} dt \quad \text{............................} \quad (2.46)$$

and

$$f(t) = \int_{-\infty}^{+\infty} \frac{d\omega}{2\pi} \left[\int_{-\infty}^{+\infty} f(t)e^{-j\omega t} dt \right] e^{j\omega t} \quad \text{...............} \quad (2.47)$$

The inner part of the two integrations is what we denote by $F(\omega)$ and we call it Fourier transform of $f(t)$. That is: -

$$F(\omega) = \int_{-\infty}^{+\infty} f(t)e^{-j\omega t} dt \quad \text{............................} \quad (2.48)$$

In a similar fashion, we have the inverse Fourier transform of $F(\omega)$ that is applied to synthesize $f(t)$. This is given by the expression in Eq. (2.49): -

$$f(t) = \frac{1}{2\pi} \int_{-\infty}^{+\infty} F(\omega)e^{j\omega t} d\omega \quad \text{............................} \quad (2.49)$$

In the case just discussed, it must be noted that $f(t)$ represents the expression for a single pulse or transient only.

Let us take another **example of a a** rectangular pulse and let us consider a rectangular pulse, $f(t)$ in Fig. 2.16, as given in Eq. (2.50):

$$f(t) = \begin{cases} 1, -\tau/2 < t < +\tau/2 \\ 0, Otherwise \end{cases} \quad \text{.............................} \quad (2.50)$$

The Fourier transform of the $f(t)$ in Eq. (2.50) is given as $F(\omega)$ expressed as in Eq. (2.51):

$$F(\omega) = \int_{-\infty}^{-\tau/2} f(t)e^{-j\omega t}\,dt + \int_{-\tau/2}^{+\tau/2} f(t)e^{-j\omega t}\,dt + \int_{+\tau/2}^{+\infty} f(t)e^{-j\omega t}\,dt \dots (2.51)$$

$$= 0 + \int_{-\tau/2}^{+\tau/2} f(t)e^{-j\omega t}\,dt + 0 = \left[\frac{e^{-j\omega t}}{-j\omega}\right]_{-\tau/2}^{+\tau/2} = \frac{1}{j\omega}\left(e^{j\omega\tau/2} - e^{-j\omega\tau/2}\right)$$

$$= \frac{2}{\omega}Sin\left(\frac{\omega\tau}{2}\right) = \tau\frac{Sin(\omega\tau/2)}{\omega\tau/2}$$

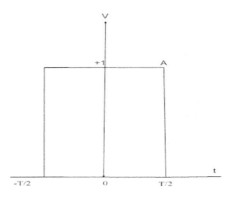

Fig. 2.16. A rectangular pulse of width τ

With slight re-arrangement, we get Eq. (2.52):

$$\frac{F(\omega)}{\tau} = \frac{Sin(x)}{x} \quad\text{--}\quad (2.52)$$

where

$$x = \frac{\omega\tau}{2} \quad\text{--}\quad (2.53)$$

2.8 Fourier transform theorems

2.8.1 Shift theorem

The Fourier transform of any time function, $f(t)$ delayed by τ_0 is simply the same Fourier transform delayed by a phase factor $e^{-j\omega\tau_0}$. This can be proved as an assignment to the reader.

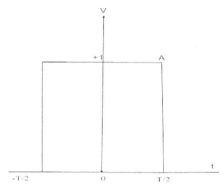

Fig. 2.17. $\dfrac{F(\omega)}{\tau}$ as a trace of $\dfrac{Sin(x)}{x}$ curve

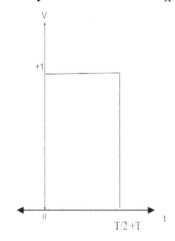

Fig. 2.18. $F(\omega) = \tau\dfrac{Sin(x)}{x}e^{-j\omega\tau_0}$

Fig. 2.17 and Fig. 2.18 respectively show $\dfrac{F(\omega)}{\tau}$ as a trace of

$\dfrac{Sin(x)}{x}$ curve and the trace of $F(\omega) = \tau \dfrac{Sin(x)}{x} e^{-j\omega\tau_0}$.

2.8.2 Impulse theorem

Certain time-functions can be reduced to a set of impulses by repeated differentiation.

If $F(\omega)$ is the Fourier transform of f(t), then differentiating f(t) *n* times gives Eq. (2.54).

$$\frac{d^n}{dt^n}[f(t)] = \frac{d^n}{dt^n} \frac{1}{2\pi} \int_{-\infty}^{+\infty} F(\omega)e^{j\omega t} d\omega \quad\quad\quad (2.54)$$

$$= (j\omega)^n \frac{1}{2\pi} \int_{-\infty}^{+\infty} F(\omega)e^{j\omega t} d\omega \quad\quad\quad (2.55)$$

This implies that Eq. (2.54) can lead to Eq. (2.56).

$$£\left[\frac{d^n}{dt^n}[f(t)]\right] = (j\omega)^n F(\omega) \quad\quad\quad (2.56)$$

2.8.3 Convolution theorem

Suppose that $v_1(t)$ has the Fourier transform $V_1(f)$ and $v_2(t)$ has the Fourier transform $V_2(f)$, we can systematically identify $v(t)$ whose Fourier transform is $V_1(f)V_2(f)$. The theorem that is composed of convolution integrals dictates that:

$$v(t) = \int_{-\infty}^{+\infty} v_1(\tau)v_2(t-\tau)d\tau \quad\quad\quad (2.57)$$

or

$$v(t) = \int_{-\infty}^{+\infty} v_2(\tau) v_1(t - \tau) d\tau \quad\text{(2.58)}$$

Eqs. (2.57) and (2.58) represent convolution integrals.

2.8.3 Perseval's theorem

Consider Eqs. (2.59) and (2.60)which mathematically represent the Perseval's.

$$v(t) = \int_{-\infty}^{+\infty} V(f) V^*(f) df \quad\text{(2.59)}$$

$$= \int_{-\infty}^{+\infty} |V(f)|^2 df = \int_{-\infty}^{+\infty} [v(t)]^2 dt \quad\text{(2.60)}$$

The theorem states that "the energy may be written as the superposition of energy due to individual spectral components separately."

2.8.4 Time and frequency relations

More compactly, we denote the transform pair $\left(v(t) \leftrightarrow V(f)\right)$ to indicate a signal and its transform or *spectrum*. Much translated as in Eq. (2.61) and Eq. (2.62): -

$$V(f) = \pounds[v(t)] \quad\text{(2.61)}$$

$$v(t) = \pounds^{-1}[V(f)] \quad\text{(2.62)}$$

2.9 Theorems and transform pairs

2.9.1 Superposition

This theorem best applies to the Fourier transform in the sense that, if a_1 and a_2 are constants and

$$v(t) = a_1 v_1(t) + a_2 v_2(t) \quad\text{(2.63)}$$

then

$$\pounds[v(t)] = a_1 \pounds[v_1(t)] + a_2 \pounds[v_2(t)] \quad\text{(2.64)}$$

With an arbitrary number of terms, the *superposition* or *linearity* theorem can be written as in Eq. (2.65) as:

$$\sum_k a_k v_k(t) \leftrightarrow \sum_k a_k V_k(f) \quad\text{(2.65)}$$

The theorem states that "Linear combinations in time domain become linear combinations in frequency domain." Significance-wise, the theorem facilitates spectral analysis when the signal in question is a linear combination of functions whose individual spectra are known.

2.9.2 Time delay and scale change

Given a time function, $v(t)$, various other waveforms can be generated from it by modifying the argument of the function. Replacing t by $(t - t_d)$, we get a time function $v(t - t_d)$ which is a time-delayed signal. It is to be noted that $v(t - t_d)$ will have the same shape as $v(t)$ but shifted t_d units to the right along the time axis. In the frequency domain, time delay causes an added linear phase with slope $-2\pi t_d$. This means that the Fourier transform pair will be represented as in Eq. (2.66):

$$v(t-t_d) \leftrightarrow V(f)e^{-j2\pi f t_d} \quad \text{---} \quad (2.66)$$

Something important to note down, though, is that the amplitude spectrum will remain unchanged. This reflects that: -

$$\left| V(f)e^{-j2\pi f t_d} \right| = \left| V(f) \right| \left| e^{-j2\pi f t_d} \right| = \left| V(f) \right| \quad \text{-----------------------} \quad (2.67)$$

The proof follows by starting with the Fourier transform pair $(V(f) \leftrightarrow v(t))$ and identifying the pair-mate of $v(t-t_d)$.

If we set $\lambda = t - t_d$ and use the expression $2\pi f = \omega$, which gives an obvious equation, $dt = d\lambda$, we can then find the transform as follows in Eq. (2.68):

$$\pounds[v(t-t_d)] = \int_{-\infty}^{+\infty} v(t-t_d)e^{-j\omega t} dt \quad \text{---} \quad (2.68)$$

$$= \int_{-\infty}^{+\infty} v(\lambda)e^{-j\omega(\lambda+t_d)} d\lambda$$

$$= \left[\int_{-\infty}^{+\infty} v(\lambda)e^{-j\omega\lambda} d\lambda \right] e^{-j\omega t_d}$$

$$= V(f)e^{-j\omega t_d}$$

2.9.3 Scale change

In our traditional $v(t)$, if we replace t with αt, we end up with a horizontally scaled image of $v(t)$. Two basic cases exist: -

(i) A function $v(\alpha t)$ is said to be expanded if $|\alpha| < 1$

(ii) A function $v(\alpha t)$ is said to be compressed if $|\alpha| > 1$

Negative α implies time reversal and expansion or compression. Practically, these effects may occur during playback of recorded signals, For example: -

$$v(\alpha t) \leftrightarrow \frac{1}{|\alpha|}V\left(\frac{f}{\alpha}\right), \text{ for } \alpha \neq 0 \quad\quad (2.69)$$

Compressing a signal expands its spectrum, and vice versa. If $\alpha = -1$, then $v(-t) \leftrightarrow V(-f)$. To prove the just presented facts, let us first consider for the case of $\alpha = 0$. This tells us that $\alpha = -|\alpha|$. Now, let us set $\lambda = -|\alpha|t$. This implies that $dt = \frac{-d\lambda}{|\alpha|}$. Therefore, we can come up with the following substitution:-

$$£\left[v(-|\alpha|t)\right] = \int_{-\infty}^{+\infty} v(-|\alpha|t)e^{-j\alpha t}dt \quad\quad (2.70)$$

$$= \frac{-1}{|\alpha|}\int_{+\infty}^{-\infty} v(\lambda)e^{-j\omega\frac{\lambda}{\alpha}}d\lambda$$

$$= \frac{1}{|\alpha|}\int_{-\infty}^{+\infty} v(\lambda)e^{-j2\pi(f/\alpha)\lambda}d\lambda$$

$$= \frac{1}{|\alpha|}V\left(\frac{f}{\alpha}\right)$$

It is to be noted that the use has been done of the general relationship

$$\int_{a}^{b} x(\lambda)d(-\lambda) = -\int_{-a}^{-b} x(\lambda)d(\lambda) = \int_{-b}^{-a} x(\lambda)d(\lambda) \quad\quad (2.71)$$

2.9.4 Frequency translation and modulation

The transform theorems can be generated by the use of the duality of the time-delay theorem given in Eq. (2.72):

$$v(t)e^{j\omega_c t} \leftrightarrow V(f - f_c), \text{ where } \omega_c = 2\pi f_c \quad\quad (2.72)$$

This is designated as frequency translation of complex modulation.

As an example, let us consider the band-limited spectrum as in Fig. 2.19 with $v(t) \rightarrow V(f)$ and Fig. 2.19. showing the transformed band-limited spectrum with $v(t)e^{j\omega_c t} \rightarrow V(f - f_c)$.

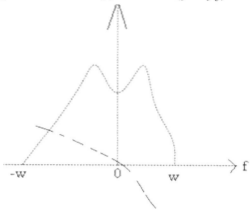

Fig. 2.18. Band limited spectrum with $v(t) \rightarrow V(f)$

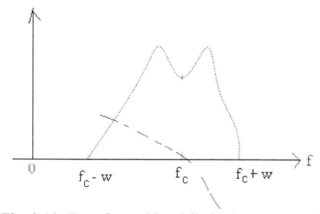

Fig. 2.19. Transformed band-limited spectrum with
$$v(t)e^{j\omega_c t} \rightarrow V(f - f_c)$$

The function $v(t)e^{j\omega_c t}$ might appear somewhat very theoretical since it does not represent a very real time practical function and can not occur as a telecommunications signal. Nevertheless, signals of the form $v(t)Cos(\omega_c t + \phi)$ are common.

2.9.5 Modulation theorem

The transform pair of $v(t)Cos(\omega_c t + \phi)$ can be realized to Eq. (2.73).

$$\frac{e^{j\phi}}{2}V(f - f_c) + \frac{e^{-j\phi}}{2}V(f + f_c) \quad\text{(2.73)}$$

The above observation complies with the fact that, by multiplying a signal by a sinusoid, translates its spectrum *up* and *down* in frequency by f_c.

2.9.6 Differentiation theorem

Something to be clear before applying this theorem is that, the theorem should not be applied before checking to make sure that the differentiated or integrated signal is Fourier transformable.

Starting from Eq. (2.74):

$$v(t) = \int_{-\infty}^{+\infty} V(f)e^{j2\pi ft}\,df \quad\text{(2.74)}$$

Differentiating both sides with respect to t, we can have Eq. (2.75);

$$\frac{d}{dt}v(t) = \frac{d}{dt}\left[\int_{-\infty}^{+\infty} V(f)e^{j2\pi ft}\,df\right] \quad\text{(2.75)}$$

$$= \int_{-\infty}^{+\infty} V(f)\left(\frac{d}{dt}e^{j2\pi ft}\right)df$$

$$= \int_{-\infty}^{+\infty}\left[j2\pi f V(f)\right]e^{j2\pi ft}\,df$$

The above observation implies that $\left(\dfrac{d}{dt}v(t), j2\pi f V(f)\right)$ is a Fourier transform pair. That means:

$$\frac{d}{dt}v(t) \leftrightarrow j2\pi f V(f) \quad \text{---} \quad (2.76)$$

Similarly, by iteration, we can generalize the differentiation theorem by the expression in Eq. (2.77):

$$\frac{d^n}{dt^n}v(t) \leftrightarrow (j2\pi f)^n V(f) \quad \text{---------------------------------} \quad (2.77)$$

2.9.7 Integration theorem

The theorem's statement simplifies in to Eq. (2.78):

$$If V(0) = \int_{-\infty}^{+\infty}v(\lambda)d\lambda = 0, \text{ then } \int_{-\infty}^{+\infty}v(\lambda)d\lambda \leftrightarrow \frac{1}{j2\pi f}V(f) \text{ ....} \quad (2.78)$$

$V(0) = 0$ corresponds to the zero net area condition that ensures that the integrated signal goes to zero as $t \to \infty$. The physical significance of the differentiation is that it enhances the high frequency components of a signal, since;

$$|j2\pi f V(f)| > |V(f)|, \text{ for } |f| > \frac{1}{2\pi} \quad \text{---------------------------------} \quad (2.79)$$

On the other hand, integration suppresses the high frequency components.

A dual of the differentiation theorem is as in Eq. (2.80):

$$t^n v(t) \leftrightarrow \frac{1}{(-j2\pi)^n}\frac{d^n}{df^n}V(f) \quad \text{---------------------------------} \quad (2.80)$$

It is left as an assignment to derive the above relationship for $n = 1$ by differentiating the transform integral $£[v(t)]$ with respect to f.

2.9.8 Convolution and multiplication

The mathematical operation known as *convolution* ranks high among the tools used by telecommunications engineers. There exit convolution in the time domain and convolution in the frequency domain.

For two functions of the same variables, say $v(t)$ and $w(t)$, the convolution is defined by: Eq. (2.81) as:

$$v(t)*w(t) \underline{\Delta} \int_{-\infty}^{+\infty} v(\lambda)w(t-\lambda)d\lambda = 0 \dotfill (2.81)$$

Expression in Eq. (2.81) is commonly named as convolution integral, which can be in mathematical notation represented as in Eq. (2.82):

$$v*w = [v(t)]*[w(t)] \dotfill (2.82)$$

The introduced λ acts as a dummy variable.

Taking an example with graphical representations of the functions and their convolutions from Fig. 2.20 through Fig. 2.23

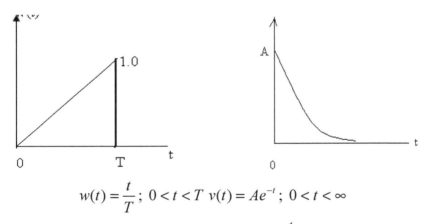

$$w(t) = \frac{t}{T}; \ 0 < t < T \quad v(t) = Ae^{-t}; \ 0 < t < \infty$$

Fig. 2.20. Graphical representation of $w(t) = \dfrac{t}{T}$ for $0 < t < T$ and for

$$v(t) = Ae^{-t} \ 0 < t < \infty$$

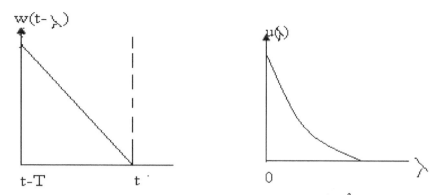

Fig. 2.21. Graphical representation of $w(t - \lambda) = \dfrac{t - \lambda}{T}$ and $v(t)=v(\lambda)$

for $0 < t - \lambda < T$

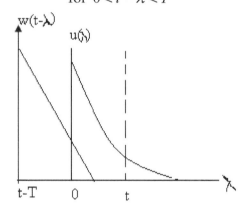

Fig. 2.22. Graphical representation of $w(t - \lambda)$ and $v(\lambda)$

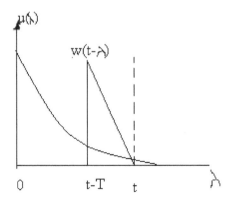

Fig. 2.23. Graphical representation of $v(\lambda)$ and $w(t-\lambda)$

The pictorial visualization of the functions has been given in Fig. 2.20 through Fig. 2.23

$$v * w(t) = \int_0^t Ae^{-\lambda}\left(\frac{t-\lambda}{T}\right) d \quad \text{............................ (2.83)}$$

$$= \frac{A}{T}\left(t - 1 + e^{-t}\right); \ 0 < t < T \quad \text{............................ (2.84)}$$

It can be noted that when $t > T$, we have Eq. (2.85)

$$v * w(t) = \int_{-T}^t Ae^{-\lambda}\left(\frac{t-\lambda}{T}\right) d\lambda \quad \text{............................ (2.85)}$$

$$= \frac{A}{T}\left(T - 1 + e^{-T}\right)e^{-(t-T)} \text{ for } t > T$$

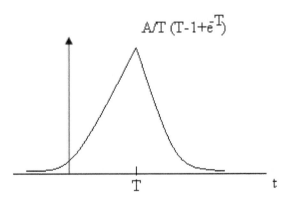

Fig. 2.24. Graphical representation of $\frac{A}{T}\left(T - 1 + e^{-T}\right)$

Fig. 2.24 concludes the result of the convolution operation.

Since the convolution is commutative, the same result must be obtained by reversing v and sliding it past w.

In general, the following properties hold for convolution theorem:-

(i)   The operation is commutative: $v * w = w * v$
(ii) The operation is associative: $v * (w * z) = (v * w) * z$
(iii) The operation is distributive: $v * (w + z) = (v * w) + (v * z)$

To summarize, let us present the two convolution theorems as in Eq. (2.86), which must be proved as an exercise: -

$$v * w(t) \leftrightarrow V(f)W(f) \text{ and } v(t)w(t) \leftrightarrow V * W(f) \dotfill (2.86)$$

The physical significance of the two operations in Eq. (2.86) is the indication that convolution in the time domain becomes multiplication in the frequency domain, while multiplication in the time domain becomes convolution in the frequency domain.

Taking example of convolving $v(t)$ and $w(t)$ as represented in Fig. 2.24 where $v(t)$ is a rectangular pulse of amplitude A_1 and duration T_1 and $w(t)$ has A_2 and T_2, respectively.

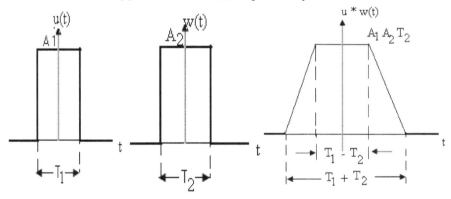

Fig. 2.24. Graphical representations of $v(t)$, $w(t)$, and $v * w(t)$

The mathematical and pictorial analyses can easily prove that the expression in Eq. (2.87) is justifiable.

$$V(f)W(f) = (A_1\tau_1 Sincf\tau_1)(A_2\tau_2 Sincf\tau_2 \text{ if } \tau_1 > \tau_2 \dotfill (2.87)$$

2.10 Power Spectral Density

2.10.1 Power Spectrum

For a periodic signal, the average power associated with a voltage $f(t)$ in a resistance of 1 ohm is given by Eq. (2.88): -

$$P_{av} = \frac{1}{T}\int_{-T/2}^{+T/2} f^2(t)dt = \frac{1}{T}\int_{-T/2}^{+T/2} f(t)\sum_{-\infty}^{+\infty} C_n e^{jn\omega t}dt \quad \text{(2.88)}$$

$$= \sum_{-\infty}^{+\infty} C_n \frac{1}{T}\int_{-T/2}^{+T/2} f(t)e^{jn\omega t}dt = \sum_{-\infty}^{+\infty} C_n C_{-n} \quad \text{(2.89)}$$

Therefore, from Eq. (2.89), we get the simplified expression for P_{av} which is given in Eq. (2.90):

$$P_{av} = \sum_{-\infty}^{+\infty} |C_n|^2 \, watts \quad \text{(2.90)}$$

For a discontinuous signal, such as a single pulse, $P_{av} \to 0$, since $\frac{1}{T} \to 0$ as $T \to \infty$. Therefore, the total energy associated with a signal in a resistance of 1 ohm becomes more meaningful. This implies that, energy, W, can be expressed in terms of P_{av} as in Eq. (2.91):

$$W = P_{av}xT = \int_{-\infty}^{+\infty} f^2(t)dt \quad \text{(2.91)}$$

But, from the fundamental definition, $f(t) = \frac{1}{2\pi}\int_{-\infty}^{+\infty} F(\omega)e^{j\omega t}d\omega$, we, hence, can have Eq. (2.92);

$$W = \int_{-\infty}^{+\infty} f(t)\frac{1}{2\pi}\int_{-\infty}^{+\infty} F(\omega)e^{j\omega t}d\omega dt \quad \text{(2.92)}$$

$$= \int_{-\infty}^{+\infty} \frac{d\omega}{2\pi} F(\omega) \int_{-\infty}^{+\infty} f(t)e^{j\omega t} dt$$

Therefore, Eq. (2.93) follows as;

$$W = \frac{1}{2\pi} \int_{-\infty}^{+\infty} F(\omega)F(-\omega)d\omega = \frac{1}{2\pi} \int_{-\infty}^{+\infty} |F(\omega)|^2 d\omega \quad \text{................} \quad (2.93)$$

We hence, realize that the Rayleigh's theorem that associates *energy spectral density*, $|F(f)|^2$, follows. The energy, W, will therefore be expressed as in Eq. (2.94);

$$W = \int_{-\infty}^{+\infty} |F(f)|^2 df \ \ joules \quad \text{..} \quad (2.94)$$

2.10.2 Power Spectral Density

Let start with the quantity of normalized power, $dS(f)$. Expressing $dS(f)$ with differentiation, we can have Eq. (2.95):

$$dS(f) = \frac{dS(f)}{df} df = G(f)df \quad \text{..} \quad (2.95)$$

where $G(f)$ is the power spectral density given by Eq. (2.96):

$$G(f) = \frac{dS(f)}{df} \quad \text{..} \quad (2.96)$$

It is clear that, for the range df, the power is $G(f)df$.

In a similar fashion, the logic shows that, the power in the positive-frequency range f_1 to f_2 is as in Eq. (2.97):-

$$S(f_1 \le f \le f_2) = \int_{f_1}^{f_2} G(f)df \quad \text{..............................} \quad (2.97)$$

The power in the negative-frequency range $-f_2$ to $-f_1$ is given in Eq. (2.98):

$$S(-f_2 \le f \le -f_1) = \int_{-f_2}^{-f_1} G(f)df \quad \text{(2.98)}$$

To sum-up, the total power in the real frequency range f_1 to f_2 is given in Eq. (2.99):

$$S(f_1 \le |f| \le f_2) = \int_{-f_2}^{-f_1} G(f)df + \int_{f_1}^{f_2} G(f)df \quad \text{(2.99)}$$

To conclude, let us see an example where power density of $\dfrac{df}{dt}$ is to be specified. It is stated that a power signal $f(t)$ has a power $S_f(\omega)$. We need to find out the power density spectrum of the signal $\dfrac{df}{dt}$.

To Solve this, we consider $\dfrac{df}{dt}$ as an output obtained if a signal $f(t)$ is passed through am ideal differentiator as shown in Fig. 2.25.

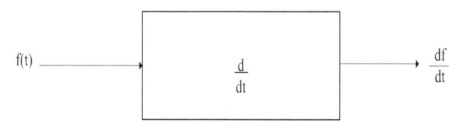

Fig. 2.26. Differentiator to get $\dfrac{df}{dt}$ from $f(t)$

However, from the *time differentiation property*, if $f(t) \leftrightarrow F(\omega)$, then $\dfrac{df}{dt} \leftrightarrow j\omega F(\omega)$. Therefore, the transfer function of the differentiation is $j\omega$. For an ideal differentiation we have Eq. (2.100):

$$|H(\omega)|^2 \leftrightarrow |j\omega|^2 = \omega^2 \quad\text{..}\quad (2.100)$$

Hence, the power density spectrum of $\dfrac{df}{dt}$ is given by Eq. (2.101) as:

$$S_{\overset{o}{f}}(\omega) = |H(\omega)|^2 S_f(\omega) = \omega^2 S_f(\omega) \quad\text{................................}\quad (2.101)$$

Therefore,

$$S_{\overset{o}{f}}(\omega) = \omega^2 S_f(\omega) \text{ is the power density of } \dfrac{df}{dt} \text{ as required.}$$

Chapter Three

Liner Continuous Wave (CW) Modulation

3.1 Introduction

Practically and in daily applications, modulation is necessary in order to transmit information. Modulation involves the phenomenon of varying some parameters of a basic electromagnetic wave (EM wave). EM wave is customary acting as, and, in fact is the carrier wave. Consider the simplest schematic operation to modulation as shown in Fig. 3.1.where the original information in a form unsuitable for distant transmission is passed to be processed to obtain information at a regular transmittable higher frequency. This process is what we refer to as modulation process.

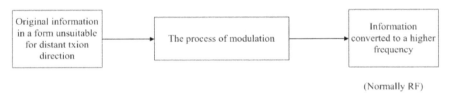

(Normally RF)

Fig. 3.1. Schematic operation to modulation

The main objective of various devised modulating methods converges in to transmitting the required information as effectively as possible with the minimum amount of distortion. While performing modulation or devising mechanisms for modulation, the following are the essential factors to be carefully considered:—(i) the signal power, (ii) the bandwidth,(iii) the distortion, and (iv)the noise power. The four factors contribute into a hybrid parameter called *signal-to-noise ratio* (SNR).

Historically and technically, modulating techniques have been competing with one another under given practical considerations and requirements. The following are the roughly grouped modulating techniques:-

(i)   Analogue methods which use Sine wave as the carrier signal
(ii)   Pulse methods which use a digital/pulse train as the carrier signal
(iii) Digital methods which use a digital data signal

It is a true fact that analogue methods and systems have largely been exploited and still in use, the capital investment in existing systems and the basic simplicity being the primary reasons, but pulse and digital methods have recently found a growing use and demand brought about by the acceptance of digital trend, globally.

3.2 Overview of the modulation methods

The fundamental modulation methods are the analogue modulation methods. There are technically two most important analogue modulation schemes. These are: -

(a)   Amplitude modulation (AM) with examples including VSB, DSBSC, and SSBSC; and.
(b)   Angle modulation—Frequency modulation (FM and Phase modulation (PM) with examples including the modulation used in VHF broadcasting, satellite communication systems, and FM radar.

The limitations of AM are basically the provision of narrowband systems and the direct effect of noise on signal amplitude. On the other side, the main limitation of angle modulation is the fact that it requires much greater bandwidth.

As, it will be discussed in the subsequent chapters and sections, the pulse methods are classified into several other sub-categories including: (i) pulse amplitude modulation (PAM); (ii) pulse duration modulation (PDM); (iii) pulse position modulation (PPM); (iv) pulse code modulation (PCM); (v) delta modulation (DM); (vi) delta PCM (DPCM); and (vii) delta sigma modulation (DSM)

For the case of the digital methods, it is easy to, approximately, correspond them to AM, FM, and PM techniques. The three traditional techniques are (i) the amplitude shift-keying (ASK); (ii) the frequency shift-keying (FSK); and (iii) the phase shift-keying (PSK).

In general, **modulation involves two waveforms**—a modulating signal that represents the message and a carrier wave that suits the particular application in a given telecommunications system under context. A modulator systematically alters the carrier wave in correspondence with the variations of the modulating signals. The modulated wave carries the message information.

Let us the examples in Fig. 3.2 and Fig. 3.3 which, respectively, represent a simplest form of modulating signal and a modulated waveform.

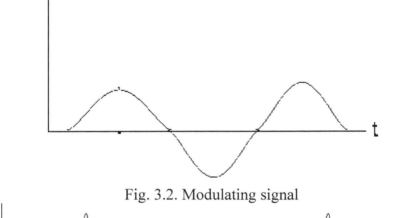

Fig. 3.2. Modulating signal

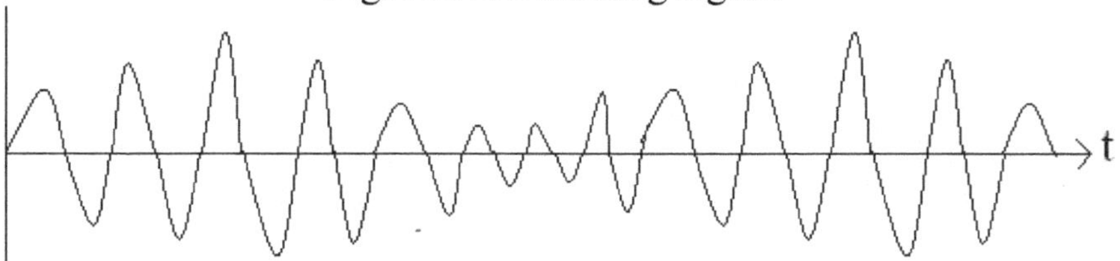

Fig. 3.3. Modulated waveform

All methods for sinusoidal carrier modulation are grouped as *continuous-wave (CW) modulation.* CW modulation produces frequency translation.

It is common question asked why do we, at all, need modulation? The benefits and applications of modulation include: (i) provision for efficient transmission; (ii) guarantee to overcome hardware limitations; (iii) tendency to reduce noise and interference; (iv) manageable frequency assignment task; and (v) simplifies multiplexing.

3.3 Amplitude modulation (AM)

The process of varying the amplitude of a radio frequency (RF) carrier wave by a modulating voltage is known as *amplitude modulation (AM)*.

Let us consider an RF carrier wave, v_c, represented in Eq. (3.1) where $\omega_c = 2\pi f_c$.

$$v_c = V_c Sin(\omega_c t) \dots\dots\dots (3.1)$$

If the modulating signal, v_m, has the form of Eq. (3.2), where $\omega_m = 2\pi f_m$, then a modulating frequency f_m becomes an important parameter in transmission.

$$v_m = V_m Sin(\omega_m t) \dots\dots\dots (3.2)$$

The amplitude of the modulated carrier varies sinusoidally between the values of $(V_c + V_m)$ and $(V_c - V_m)$ as shown in Fig 3.4.

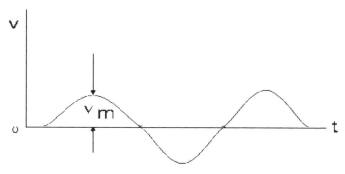

Fig 3.4. Modulating signal with peak voltage of V_m

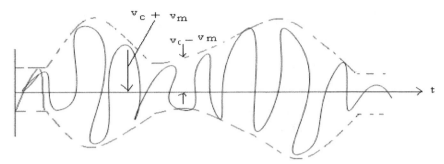

Fig 3.4. AM carrier signal

If the modulation factor is given as in Eq. (3.3) as:

$$\frac{V_m}{V_c} = m = modulation\ factor = depth\ of\ modulation \quad\text{......}\quad (3.3)$$

$$V_m = mV_c \quad\text{......}\quad (3.4)$$

then we can have the expression in Eq. (3.4) which leads to the expression for the modulated carrier given by Eq. (3.5):

$$v_c = (V_c + V_m Sin(\omega_m t))Sin(\omega_c t)$$

$$= V_c Sin(\omega_c t) + mV_c Sin(\omega_c t)Sin(\omega_m t) \quad\text{......}\quad (3.5)$$

However, from the basics, we have Eq. (3.6)

$$Sin(\omega_c t)Sin(\omega_m t) = \frac{1}{2}[Cos(\omega_c - \omega_m)t - Cos(\omega_c + \omega_m)t] \quad\text{......}\quad (3.6)$$

Therefore, v_c will be given in Eq. (3.7) as:

$$v_c = V_c Sin(\omega_c t) + \left(\frac{mV_c}{2}\right)Cos(\omega_c - \omega_m)t - \left(\frac{mV_c}{2}\right)Cos(\omega_c + \omega_m)t \quad\text{....}$$

$$\text{......}\quad (3.7)$$

3.4 The AM spectrum

From the fundamental expression for AM as given in Eq. (3.8),

$$v_c = V_c Sin(\omega_c t) + \left(\frac{mV_c}{2}\right)Cos(\omega_c - \omega_m)t - \left(\frac{mV_c}{2}\right)Cos(\omega_c + \omega_m)t \ ...$$

$$..(3.8)$$

we can see that v_c is primarily constituted with three frequency components which are on the right hand side of Eq. (3.8). The first represents the carrier frequency component, the second term corresponds to the lower-side frequency (LSF) component, and the last term signifies the upper-side frequency (USF) component. The side frequencies are above or below the carrier frequency by an amount equal to f_m.

For a complex modulating signal, like speech, numerous frequency components are produced above and below the f_c. These make the lower sidebands (LSB) and the upper sidebands (USB).

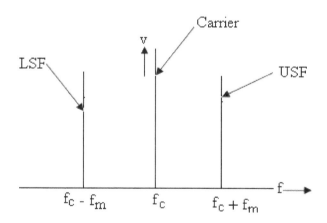

Fig 3.5. Single tone modulation

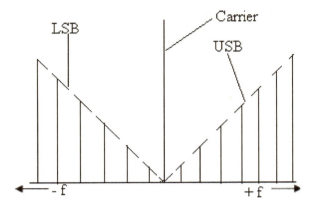

Fig 3.6. Complex tone modulation

3.4.1 Power in AM spectrum

In a 1 ohm load resistor, the average power corresponds to the square of the voltage, V^2. This is carrier power of the transmission. Thus, *the carrier power* $= V^2$. Therefore, the sideband power is given as in Eq. (3.9) as:

$$The\ sideband\ power = 2\left(\frac{mV}{2}\right)^2 = \frac{m^2V^2}{2} \quad\text{........................}\quad (3.9)$$

Therefore, the total power is computes as in Eq. (3.10) as:

$$The\ total\ power = V^2 + \frac{m^2V^2}{2} = V^2\left(1 + \frac{m^2}{2}\right) \quad\text{...................}\quad (3.10)$$

We, then, have Eq. (3.11):

$$\frac{P_{sideband}}{P_{total}} = \frac{m^2V^2/2}{V^2\left(1 + m^2/2\right)} = \frac{m^2}{2 + m^2} \quad\text{.......................................}\quad (3.11)$$

Based on the above analysis, it is to be noted that the value of sidebands power, $\left(P_{sideband}\right)_{max} = 50\%$ of the carrier power when $m = 1.0$. It must also be noted that the sideband power depends on m^2, but for practical reasons, m_{av} is chosen between 30 percent and 50 percent, such that; $30\% \leq m_{av} \leq 50\%$.

Finally, the carrier and one sideband may be suppressed without destroying the information because the information is also present in the remaining sideband.

3.4.2 AM modulators

The function of an AM modulator is to *modulate* a carrier wave which results in sum and difference frequencies, together with the carrier. This may be achieved by using valves or transistors operating as *non-liner* or *liner* modulators.

3.4.2.1 Non-liner modulators

None-linear devices like semiconductor diodes and transistors may be used.

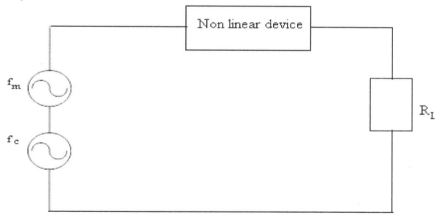

Fig 3.7. The basic circuit of AM modulator

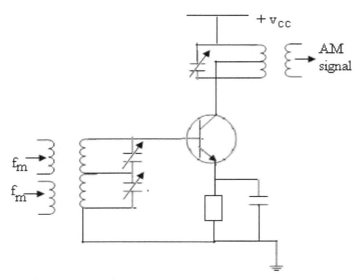

Fig 3.8. Practical arrangement of AM modulator

Fig 3.7 shows the basic circuit of AM modulator while Fig 3.8 represents a practical arrangement of AM modulator.

Let us consider that the device has a characteristic of the form of Eq.(3.12):

$$i = a + bv + cv^2 \quad\text{(3.12)}$$

where $v = input\ voltage$ and $i = output\ current$. From our fundamental AM function, we can have the validity of Eq. (3.13),

$$v = V_c Sin(\omega_c t) + V_m Sin(\omega_m t) \quad\text{(3.13)}$$

Therefore, we can expand Eq. (3.12) into Eq. (3.14) and Eq. (3.15) as:

$$i = a + b(V_c Sin(\omega_c t) + V_m Sin(\omega_m t)) + c(V_c Sin(\omega_c t) + V_m Sin(\omega_m t))^2 \quad\text{(3.14)}$$

$$= a + b V_c Sin(\omega_c t) + b V_m Sin(\omega_m t)$$
$$+ c V_c V_m Cos(\omega_c - \omega_m t) - c V_c V_m Cos(\omega_c + \omega_m)t \quad\text{(3.15)}$$
$$+ c V_c^2 Sin^2(\omega_c t) + c V_m^2 Sin^2(\omega_m t)$$

If R_L corresponds to the tuned load, it implies that the carrier sideband components can be selected to give the required AM output of the system.

3.4.2.2 Linear modulators

When good linearity is required, like the way it must be in large power transmitters, anode modulation is usually employed.

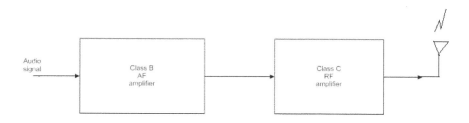

Fig 3.9. Linear modulator

In Fig 3.9, Class B AF amplifier drives Class C RF amplifier while C is biased well beyond cut-off and anode current flows in pulses for a part of the RF cycle. A fairly undistorted AM signal may be obtained with an efficiency of about 80%.

3.4.3 Double sideband suppressed carrier (DSBSC) system

The most common and simplest AM system is the double-sideband transmission where both sets of sidebands plus the carrier are transmitted. The receiver design seems to be simple and it reduces costs.

Since the carrier does not contain the information, it may be suppressed wholly or partly at the transmitter, yielding a DSBSC system. This leads into a saving in carrier power up to 96% for m=0.3. Also, the carrier is re-inserted at the receiver leading to recovery of the original modulation. The technique raises complication because of the stringent phase requirements.

For correct phasing, the modulation is present, but for a 90^0 phase error, the modulation is lost as no amplitude modulation occurs, but only phase modulation. It must be noted that PM can not be detected by an AM receiver.

The circuits used to generate DSBSC follow the principle of balanced modulation. Balanced modulator employs non-linear devices, such as valves or transistors.

Let us consider Fig 3.10 which shows a schematic balanced modulator where f_c is in phase in both halves of the circuit, but f_m is out of phase.

With k as the constant of proportionality, we can have Eqs. (3.16) and (3.16) as:

$$i_1 = k(1 + mSin(\omega_m t))Sin(\omega_c t) \text{------------------------------} (3.16)$$

and

$$i_2 = k(1 - mSin(\omega_m t))Sin(\omega_c t) \text{------------------------------} (3.17)$$

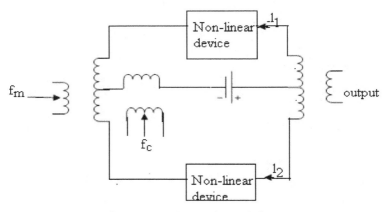

Fig 3.10.Balanced modulator

In the output transformer we have Eq. (3.18) give as:

$$i_o = |i_1 - i_2| \text{--} (3.18)$$

This means that $|i_1 - i_2|$ can be expressed in Eqs. (3.19) and (3.20) as:

$$|i_1 - i_2| = 2kmSin(\omega_m t))Sin(\omega_c t) \dots\dots\dots (3.19)$$

$$= km[Cos(\omega_c - \omega_m)t - Cos(\omega_c + \omega_m)t] \dots\dots (3.20)$$

The above expression implies that there in no carrier component present.

3.4.3.1 Cowan modulator

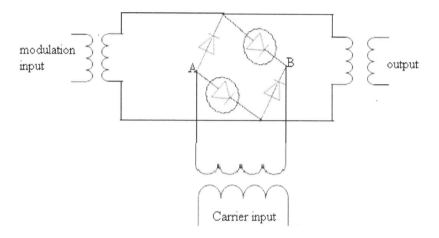

Fig 3.11.Cowan modulator

When A is positive with respect to B, we have diodes conducting and, hence, the bridge experiences a short circuit across the network. Therefore, no audio signal gets through.

When A is negative with respect to B, we have diodes reverse biased and, hence, there is a short circuit across the network. That means audio signal gets through. Therefore, the carrier wave has a square-wave switching action on the audio signal.

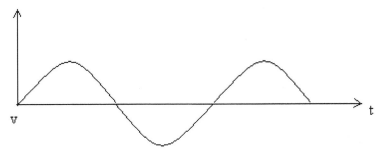

Fig 3.12.Modulating signal

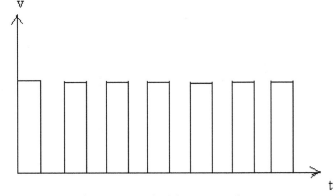

Fig 3.13.Switching waveform

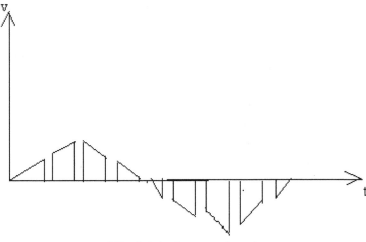

Fig 3.14. Output signal

The output signal signifies the product of the square wave function and the audio signal.

From the figures, Figs 3.12, 13, and 14 where modulating signal, switching waveform, and the output signal have been shown, it can be visualized that the square-wave switching function is given by Eq. (3.21) as:

$$v_s = \frac{1}{2} + \frac{2}{\pi}\left[Sin(\omega_c t) + \frac{1}{3}Sin3(\omega_c t) + \frac{1}{5}Sin5(\omega_c t) + ...\right] \text{------ (3.21)}$$

If the audio signal, v_m, is given by Eq. (3.22), then the output v_o is given by Eq. (3.23) as:

$$v_o = kv_s v_m = kV_m Sin(\omega_m t)\left[\frac{1}{2} + \frac{2}{\pi}\{Sin(\omega_c t) + ...\}\right] \text{------------ (3.22)}$$

$$v_m = V_m Sin(\omega_m t) \text{-- (3.23)}$$

Therefore, with k a constant wit dimension $Volt^{-1}$, we can further express Eq, (3.22) as in Eq. (3.24):

$$v_o = \frac{kV_m Sin(\omega_m t)}{2} + \frac{kV_m}{\pi}\left[Cos(\omega_c - \omega_m)t - Cos(\omega_c + \omega_m)t\right] + ... ;$$

$$\text{-- (3.24)}$$

Eq. (3.24) signifies that the output contains the audio signal and the two side-frequencies, but the carrier is absent.

3.4.3.2 Ring modulator

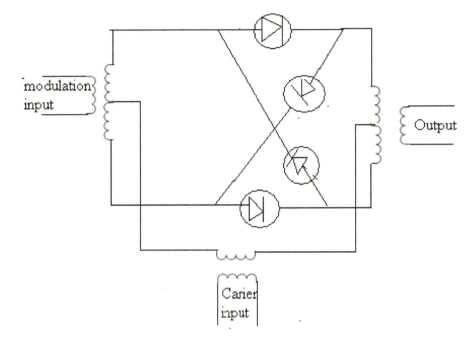

Fig 3.15. Ring modulator

The ring symbolizes an arrangement of rectifiers where a double-sided square-wave switching action is experienced on the network during positive and negative excursions of the carrier signals. We have a reversing switch operating at the f_c.

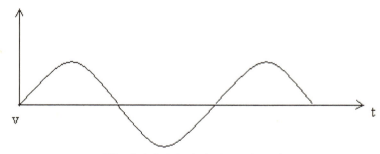

Fig 3.16. Modulating signal

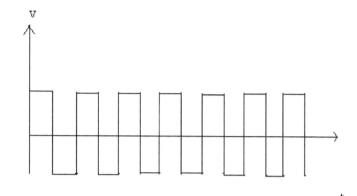

Fig 3.17. Switching waveform

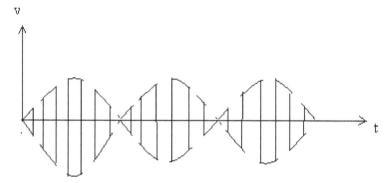

Fig 3.18. Output signal

Again if we consider the figures, Figs 3.16, through 3.18 where modulating signal, switching waveform, and the output signal have been, respectively, shown, it can be visualized that the square-wave switching function and v_m are respectively given by Eq. (3.25) and (3.26) as:

$$v_s = \frac{4}{\pi}\left[Sin(\omega_c t) + \frac{1}{3}Sin3(\omega_c t) + \frac{1}{5}Sin5(\omega_c t) + ...\right] \text{------------ (3.25)}$$

and

$$v_m = V_m Sin(\omega_m t) \dashrightarrow (3.26)$$

Then, we can express v_o as in Eqs. (3.27) and (3.28):

$$v_o = kv_s.v_m = kV_m Sin(\omega_m t).\frac{4}{\pi}\left[Sin(\omega_c t)+\frac{1}{3}Sin3(\omega_c t)+...\right] \quad (3.27)$$

$$= kv_s.v_m = 2\frac{kV_m}{\pi}\left[Cos(\omega_c - \omega_m)t - Cos(\omega_c + \omega_m)t +...\right] \quad (3.28)$$

This implies that the output contains mainly the pair of side-frequencies, but no carrier or audio frequencies are present.

3.4.4 Single sideband suppressed carrier (SSBSC) system

The information is present in either sideband. In telecommunications, the need is always there to save two important commodities of power and bandwidth. The saving can be obtained by suppressing either the lower or the upper sideband. For $m = 0.3$, as in the previous typical example, we can have up to 2% saving in power corresponding to one side-band which leads to up to 50% saving in bandwidth which is one half of the spectrum of the AM system.

3.4.4.1 How to produce an SSBSC Signal?

By obtaining DSBSC signal using a balanced modulator then to suppress the unwanted sideband by a sideband filter as shown in Fig. 3.19. The filter output leads to just one sideband.

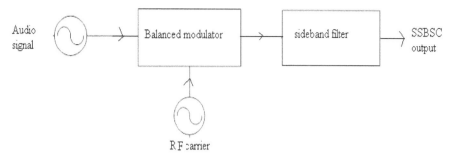

Fig 3.19. SSBSC output signal generation

Difficulties that can be encountered includes designing a filter with sharp-cut-off on either side. A narrower bandwidth leads to loss of some sideband components while increased bandwidth leads to some of the other sideband components to leak through.

Therefore, vestigial sideband transmission (VSB), which is largely applicable in TV, leads to increased bandwidth to include a *dc* signal.

(a) Phase-shift method

SSBSC signal signifies two DSBSC signals with their carrier and modulation 90^0 out of phase. Therefore, we apply two balanced modulators as shown in Fig 3.20.

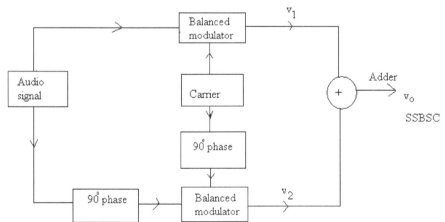

Fig 3.20. Phase-shift modulator—two balanced modulators

If v_1 and v_2 are the output voltages of the modulators, respectively given as in Eq. (3.29) and Eq. (3.30):

$$v_1 = \frac{mV_c}{2}\left[Cos(\omega_c - \omega_m)t - Cos(\omega_c + \omega_m)t\right] \quad\text{.................}\quad (3.29)$$

and

$$v_2 = \frac{mV_c}{2}\left[Cos\left\{(\omega_c t + \pi/2) - (\omega_m t + \pi/2)\right\} - Cos\left\{(\omega_c t + \pi/2) + (\omega_m t + \pi/2)\right\}\right]$$

$$\text{...} \quad (3.30)$$

$$= \frac{mV_c}{2}\left[Cos(\omega_c - \omega_m)t + Cos(\omega_c + \omega_m)t\right] \quad\text{..............}\quad (3.31)$$

Since v_o can be expressed in terms of v_1 and v_2, we will have Eq. (3.32) as:

$$v_o = v_1 + v_2 = mV_c\left[Cos(\omega_c - \omega_m)t\right] \quad\text{...........................}\quad (3.32)$$

The expression in Eq. (3.32) implies the presence of just the lower sideband.

3.4.5 Vestigial sideband (VSB) system

For a 625-line TV system, the video signal bandwidth is about 6MHz. This implies that a modulation bandwidth of 12MHz will be needed when AM system is used. For the purpose of economizing the bandwidth, double sideband (DSB) is used where a vestige of one sideband is transmitted along with the whole of the other sideband. This reduces the overall bandwidth to about 8MHz.

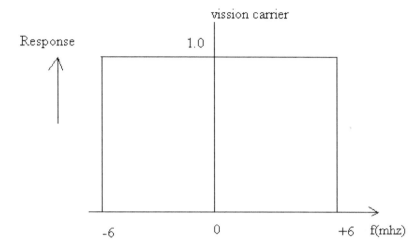

Fig 3.21.DSB characteristic

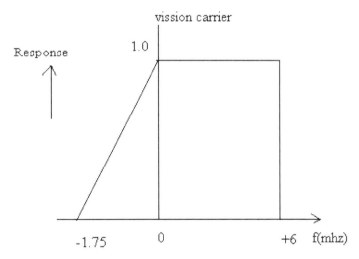

Fig 3.22. VSB transmitter characteristic

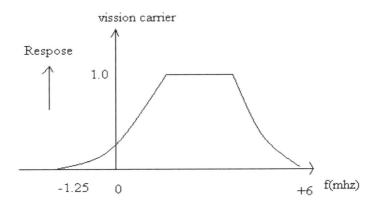

Fig 3.23. VSB receiver characteristic

Let us consider Figs 3.21, 22, and 23 which respectively show the DSB characteristic, the VSB transmitter characteristic and the VSB receiver characteristic. Transmission of the *dc* signal represents the average brightness of the picture and it is important picture detail.

Next, in the coming section, we will discuss the "Third" method of generating SSB, developed by Weaver as a means of retaining the advantage of the phase-shift method.

3.4.6 The "Third" method of generating SSB

The method was developed as a means of retaining the advantages of the phase-shift method, that is, ability to generate SSB at any frequency and use of low AF without the associated disadvantage of an AF phase-shift network required to operate over a large range of audio frequencies.

The "Third" method is in direct competition with the filter method, but it is very complex and not often used commercially.

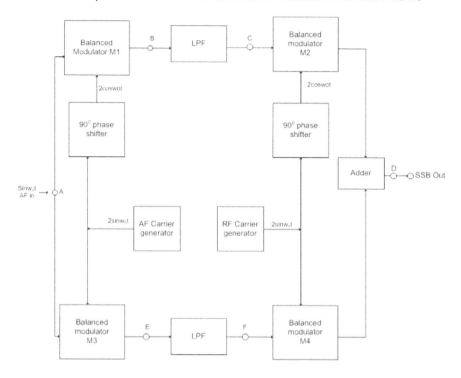

Fig 3.24. The "Third" method of generating SSB

As it can be seen in Fig 3.24, instead of trying to phase-shift the whole range of AF, this method combines them with an AF carrier f_o, which is fixed frequency in the middle of audio band. A phase-shift is then applied in this frequency only.

3.4.7 Independent sideband (ISB) system

In certain applications, both sidebands of a DSB system are used, but each sideband carries a different message. The signal may be transmitted with or without a pilot carrier. This is called *independent sideband* (ISB) transmission.

The process involved, as described in Fig 3.25, is to produce two separate SSB signals—one message, *channel one,* is superimposed on one of the sidebands and another message superimposed on the other sideband. The sidebands are arranged to occupy a 6KHz bandwidth on either side of the carrier frequency.

The two independent sidebands are then mixed with or without a carrier in a hybrid circuit.

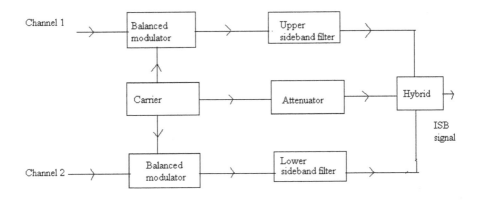

Fig 3.25. ISB system

3.5 AM Transmitter

Fig 3.26 shows a typical schematic arrangement of an AM transmitter commonly used in practice.

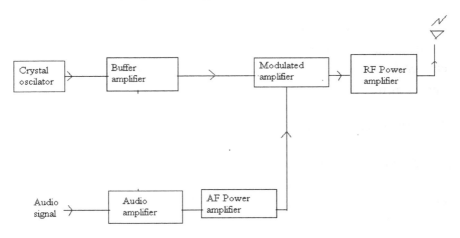

Fig 3.26. A typical schematic arrangement of an AM transmitter

In Fig. 3.26, it has been shown that the RF carrier wave is obtained from a stable crystal oscillator. A buffer amplifier ensures high stability. The modulated carrier is amplified in a high-power linear RF amplifier prior to being fed to the aerial for transmission. That means a suitable *dc* power supply is also required to provide the energy for the transmitter.

A typical medium-wave transmitter operates at a power level of about 50kW for broadcasting speech and music.

To understand the essence of AM transmission and the power delivered, let us have an example. We have an AM transmitter that has an anode modulated class-C output stage in which an audio-frequency Sine wave of 3kV peak value is developed across the secondary of the modulating transformer in series with the 5kV HT supply. The stage has an anode efficiency of 75% and delivers 1.5kW of carrier power into the tank circuit. Assuming that the RF output varies in proportion to the anode voltage and that there are no losses in the modulating transformer, we need to calculate:

(a) the depth of modulation,
(b) the mean anode current,
(c) the power supplied by the modulator, and
(d) the total RF power delivered to the tank circuit.

$$\text{The depth of modulation} = \frac{(audio)_{peak}}{(RF)_{peak}} = m = \frac{3x10^3}{5x10^3} = 0.6$$

(a) To get the anode current, I_{mean}, we must first find the *dc* power

$$the\ dc\ power = \frac{1.5x10^3}{0.75}$$

$$= 2x10^3 W = I_{mean} x V_{dc}$$

Therefore,

$$I_{mean} = \frac{2x10^3}{5x10^3} = 0.4A$$

$$\text{Power supplied by modulator} = \frac{(power)_{sideband}}{efficiency}$$

But $(power)_{sideband} = \dfrac{m^2}{2}.P_c = \dfrac{(0.6)^2}{2}x1.5x10^3 = 0.27kW$

This gives the modulation power $= \dfrac{0.27kW}{0.75} = 0.36kW$

(b) The total power, $(power)_{total} = (power)_{carrier} + (power)_{sideband}$

This implies that, $(power)_{sup\,plied} = 1.5x10^3W + 0.27kW = 1.77kW$

Therefore, the RF power supplied $= 1.77kW$.

3.6 AM detection/demodulation

For extracting or recovering the information from a modulated signal, demodulation or detection process is done. Modulated signal can be of the analogue or digital type. The demodulation techniques used can be different in each case. Demodulation of AM and FM signals require different methods. In general, linear demodulation is essential in both AM and FM receivers in order to minimize signal distortion.

3.6.1 AM detectors

To detect an AM signal, a non-linear or linear detector is required. A typical non-linear detector is a *square-law detector* and typical linear detectors are *non-coherent/envelope detector* and *coherent/synchronous detector.*

The envelope detector is the most common type used for AM signals. It is simple and it applies equally well to speech, music, or video signals while coherent detector is more employed for DSBSC or SSBSC signals and it is more critical in its operation. It depends upon exact carrier synchronization for the proper operation.

Square-law detector finds application in radar systems where waveform distortion is of no consequence. The main concern is to detect a weak pulse signal reflected from a distant target.

3.6.1.1 Super-heterodyne principle

A principle used by receiving system for an AM signal. Fig 3.27 shows the operation taking place in a super-heterodyne system.

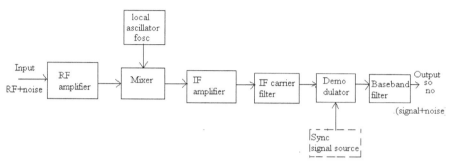

Fig 3.27. Super-heterodyne system

For the working principle, we assume that the signal has suffered great attenuation during transmission. This means that, it is in need of amplification. We also assume that the carrier of the received signal is called RF, with frequency f_{rf}.

In the mixer, the RF carrier is mixed, or multiplexed, with a sinusoidal waveform generated by a local oscillator with frequency f_{osc}. Therefore, the process of mixing is also called *heterodyning.* The f_{osc} is selected to be above the f_{rf}. This is what produces a *super-heterodyne system.*

Mixing brings sum and difference frequencies with output consisting of a carrier of $(f_{osc} + f_{rf})$ and a carrier of $(f_{osc} - f_{rf})$. The carrier of $(f_{osc} + f_{rf})$ is rejected by a filter while the carrier of $(f_{osc} - f_{rf})$ is called the *intermediate frequency (IF)* carrier.

Therefore we can have an expression for f_{if} given as in Eq. (3.33),

$$f_{if} = f_{osc} - f_{rf})\ \text{..} (3.33)$$

The modulated RF is replaced by modulated IF carrier and this phenomenon is what we call *conversion.* The mixer and the local oscillator (LO) do the conversion and act as the first detector while the demodulator acts as the second detector.

3.6.1.2 Advantages of the super-heterodyne principle

A signal at the receiver may have a power as low as some tens of pW. The required output signal may be of the order of tens of watts. Therefore, a large magnitude is required. The largest part of the required gain is provided by the IF amplifier. The critical filtering is done by the IF filter. RF amplification is employed whenever the incoming signal is very small. RF amplifiers are *low noise* devices. Finally, it can be noted that, the mixer alone provides relatively *little gain* and generates a relatively *large nose* power.

3.6.1.3 Super-heterodyne principle and multiplexing

For tuning the receiver to more than one different signal, each using a different RF carrier, we apply multiplexing technique. If we were not to take advantage of the super-heterodyne principle, we would require a receiver in which many stages of RF amplification were employed. Bad enough, each stage would be requiring tuning.

In a super-heterodyned receiver, we need but change the frequency of the oscillator to go from one RF carrier frequency to another. The reason for selecting f_{osc} higher than f_{rf} is that, within this higher selection, the fractional change in f_{osc} required to accommodate a given range of RF frequencies is smaller than would be the case for the alternative selection.

Chapter Four

Exponential Continuous Wave (CW) Modulation

4.1 Introduction

Before we embark deeper into the principles of the exponential continuous wave modulation, it is a wise ides to learn about the common properties of linear CW modulation. It is to be understood that in linear CW, the modulation spectrum is basically the translated message spectrum while the transmission bandwidth never exceeds twice the message bandwidth. Further, it must be clear that the destination signal-to-noise ratio $(S/R)_D$ is no better than the baseband transmission and it can be improved only by increasing the transmitted power.

In fact, exponential modulation differs on all three counts. In exponential modulation, the modulation spectrum is not related in a simple fashion to the message spectrum and the transmission bandwidth is usually much greater than twice the message bandwidth. Finally, exponential continuous wave modulation can provide increased SNR without increased transmitted power. This means that there is a bandwidth-power trade-off in the design of a telecommunications system.

Having realized the said three basic differences, we must realize the two basic types of exponential modulation—the phase modulation (PM) and the frequency modulation (FM).

4.2 PM and FM signals

With $\theta_c(t) = \omega_c t + \phi(t)$, let a CW signal with constant envelope but time-varying phase, such that it can be represented by Eq. (4.1) as:

$$x_c(t) = A_c Cos[\omega_c t + \phi(t)] \dotfill (4.1)$$

$$= A_c Cos\,\theta_c(t) \quad\text{..} \quad (4.2)$$

$$= A_c R_c\left[e^{j\theta_c(t)}\right] \quad\text{..................................} \quad (4.3)$$

If $\theta_c(t)$ contains the message information, $x(t)$, the process is either *angle modulation* or *exponential modulation*. There is a non-linear relationship between $x_c(t)$ and $x(t)$. Specifically, in PM, for $\phi_\Delta \leq 180^0$ we can have a definition of $\phi(t))$ expressed as in Eq. (4.4):

$$\phi(t)\underline{\Delta}\phi_\Delta x(t) \quad\text{...} \quad (4.4)$$

Therefore, $x_c(t)$ can be expressed in terms of $x(t)$ as in Eq. (4.5):

$$x_c(t) = A_c Cos\left[\omega_c t + \phi_\Delta x(t)\right] \quad\text{.................} \quad (4.5)$$

We can observe that the instantaneous phase varies directly with the modulating signal. The constant ϕ_Δ corresponds to maximum phase shift produced by $x(t)$. To prevent phase ambiguity, we set $\phi(t)$ such that Eq. (4.6) holds.

$$-180^0 \leq \phi(t) \leq +180^0 \quad\text{............................} \quad (4.6)$$

where ϕ_Δ is the phase modulation index or the phase deviation.

The rotating-phasor diagram is as shown Fig. 4.1.

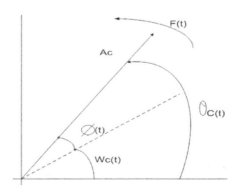

Fig. 4.1. The rotating-phasor diagram

The total angle, $\theta_c(t)$, is given as in Eq. (4.8) as:

$$\theta_c(t) = \omega_c t + \phi(t) \quad\text{(4.7)}$$

The phasor's instantaneous rate of rotation in cycles per second will be given as in Eq. (4.8):

$$f(t) \triangleq \frac{1}{2\pi} \theta_c(t) = f_c + \frac{1}{2\pi} \overset{o}{\phi}(t) \quad\text{(4.8)}$$

where

$$\overset{o}{\phi}(t) = \frac{d\phi(t)}{dt} \quad\text{(4.9)}$$

The function, $f(t)$, represents an instantaneous frequency of $x_c(t)$.

In the case of frequency modulation (FM), the instantaneous frequency of the modulated wave is defined by Eq. (4.10) as:

$$f(t) \triangleq f_c + f_\Delta x(t) \quad\text{(4.10)}$$

such that $f_\Delta < f_c$

It can be noted that $f(t)$ varies in proportion with the modulating signal and f_Δ is the frequency deviation which implies the maximum shift of $f(t)$. Therefore, the condition that $f_\Delta < f_c$ ensures that $f(t) > 0$.

Usually, it is required that $f_\Delta \ll f_c$. This condition is to preserve the band-pass nature of $x_c(t)$. Therefore, the FM wave has $\overset{o}{\phi}(t) = 2\pi f_\Delta x(t)$ and the PM implies the integration, that is Eq. (4.11):

$$\phi(t) = 2\pi f_\Delta \int_t^t x(\lambda) d\lambda + \phi(t_0) \quad\text{(4.11)}$$

OMAR FAKIH HAMAD

where $t \geq t_0$.

If $\phi(t_0) = 0$, we will have Eq. (4.12):

$$\phi(t) = 2\pi f_\Delta \int^t x(\lambda)d\lambda \text{-----------------------------} (4.12)$$

Therefore, The FM waveform is written as in Eq. (4.13):

$$x_c(t) = A_c Cos\left[\omega_c t + 2\pi f_\Delta \int^t x(\lambda)d\lambda\right] \text{----------------} (4.13)$$

Analysis shows that there is a little difference, in two ways, between PM and FM. The comparison of PM and FM is, firstly, that PM has an instantaneous phase, $\phi(t)$, as given in Eq. (4.14):

$$\phi(t) = \phi_\Delta(t)x(t) \text{-----------------------------------} (4.14)$$

while that for FM is given in Eq. (4.15) as:

$$\phi(t) = 2\pi f_\Delta \int^t x(\lambda)d\lambda \text{-----------------------------} (4.15)$$

Secondly, in terms of instantaneous frequency, $f(t)$, that for PM is given as in Eq. (4.16):

$$f(t) = f_c + \frac{1}{2\pi}\phi_\Delta \overset{o}{x}(t) \text{-----------------------} (4.16)$$

while the instantaneous frequency for FM is given as in Eq. (4.17):

$$f(t) = f_c + f_\Delta x(t) \text{-------------------------------} (4.17)$$

The amplitude of an FM or PM wave is constant. Therefore, regardless of the message $x(t)$, the average transmitted power is given by Eq. (3.18):

$$S_T = \frac{A_c^2}{2}$$ -- (4.18)

This means that the message resides in the zero crossings alone. The modulated wave does not look at all like the message waveform.

4.3 Frequency modulation (FM)

The process of varying the frequency of a carrier wave in proportion to a modulating signal is known as FM. The carrier amplitude of an FM wave is kept constant during modulation and so the power associated with an FM wave is constant. The carrier frequency increases when the modulating voltage increases positively and it decreases when the modulating voltage becomes negative.

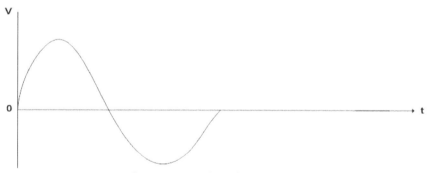

Fig. 4.2. Modulating signal

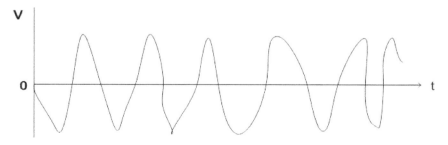

Fig. 4.3.Carrier signal

Let the let the modulating signal be represented as in Fig. 4.2 and and instantaneous carrier wave, as in Fig. 4.3, be represented by Eq. (4.19):

$$v_c = V_c Sin\omega_1 t = V_c Sin 2\pi f_1 t \quad\text{............} \quad (4.19)$$

where f_1 is the instantaneous frequency.

For positive increase in frequency, we have Eq. (4.20):

$$f_1 = f_c + \Delta f_c Sin\omega_m t \quad\text{............}\quad (4.20)$$

where f_c is the carrier frequency and Δf_c is the frequency deviation of the carrier wave due to the modulating signal of frequency f_m.

If the instantaneous carrier phase is ϕ_1, then we have Eq. (4.21):

$$\frac{1}{2\pi} \cdot \frac{d\phi_1}{dt} = f_1 = f_c + \Delta f_c Sin\omega_m t \quad\text{............}\quad (4.21)$$

This means that, Eq. (4.21) can be rearranged and be written as in Eq. (4.22):

$$\frac{d\phi_1}{dt} = 2\pi f_1 = \omega_c + 2\pi\Delta f_c Sin\omega_m t \quad\text{............}\quad (4.22)$$

This implies that:

$$\phi_1 = \omega_c t - \frac{\Delta f_c}{f_m} Cos\omega_m t \quad\text{............}\quad (4.23)$$

With $m_f = \dfrac{\Delta f_c}{f_m}$ = the modulation index, we can have ϕ_1 expressed as in Eq. (4.24):

$$\phi_1 = \omega_c t - m_f Cos\, \omega_m t \quad\text{............}\quad (4.24)$$

With $v_c = V_c Sin\phi_1$, then we have Eq. (4,25)

$$v_c = V_c Sin[\omega_c t - m_f Cos\omega_m t] \dots\dots\dots\dots\dots\dots\dots (4.25)$$

The expression in Eq. (4.25) is the FM carrier wave.

4.3.1 The FM Spectrum

From $v_c = V_c Sin[\omega_c t - m_f Cos\omega_m t]$, expanding v_c yields Eq. (4.26):

$$v_c = V_c [Sin\omega_c t Cos(m_f Cos\omega_m t) - Cos\omega_c t Sin(m_f Cos\omega_m t)] \dots (4.26)$$

From the fundamentals, we have Eq. (4.27) and Eq. (4.28), respectively given as:

$$Cos(m_f Cos\omega_m t) = J_0(m_f) - 2J_2(m_f)Cos2\omega_m t + 2J_4(m_f)Cos4\omega_m t + \dots$$
$$\dots\dots\dots\dots\dots\dots\dots\dots\dots\dots (4.27)$$

and

$$Sin(m_f Cos\omega_m t) = 2J_1(m_f)Cos\omega_m t - 2J_3(m_f)Cos3\omega_m t + \dots (4.28)$$

The coefficients $J_n(m_f)$ are Bessel functions of the first kind and order n.

Substituting $Cos(m_f Cos\omega_m t)$ and $Sin(m_f Cos\omega_m t)$ in the v_c in Eq. (4.26) we get Eq. (4.29):

$$v_c = V_c [J_0(m_f)Sin\omega_c t - J_1(m_f)\{Cos(\omega_c + \omega_m)t + Cos(\omega_c - \omega_m)t\}$$

$$- J_2(m_f)\{Sin(\omega_c + 2\omega_m)t + Sin(\omega_c - 2\omega_m)t\}$$

$$- J_3(m_f)\{Cos(\omega_c + 3\omega_m)t + Cos(\omega_c - 3\omega_m)t\} + \dots] \dots\dots\dots (4.29)$$

The phenomenon corresponds to appearance of infinite set of sidebands whose amplitudes are determined by the Bessel functions $J_1(m_f)$, $J_2(m_f)$, $J_3(m_f)$, etc.

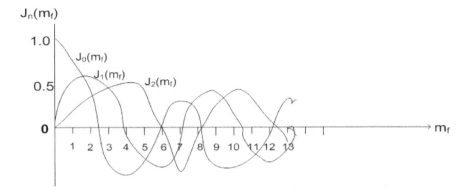

Fig. 4.4. Infinite set of sidebands whose amplitudes are determined by the Bessel functions $J_1(m_f)$, $J_2(m_f)$, $J_3(m_f)$, ..., $J_n(m_f)$.

When m_f is small, we have a few sideband frequencies of large amplitudes, and when m_f is large, we have many sideband frequencies with smaller amplitudes. In practice, it is necessary to consider a finite number of significant sideband components whose amplitudes are greater than about 4% of the un-modulated carrier.

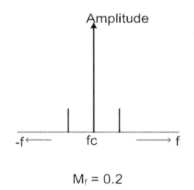

$M_f = 0.2$

Fig. 4.5. Typical plot for $m_f = 0.2$

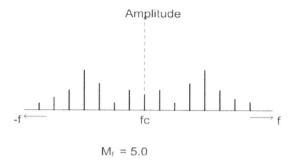

Fig. 4.6. Typical plot for $m_f = 5.0$

In practice the frequency deviation of FM system is largely determined by the available bandwidth. Most FM broadcast stations use a frequency deviation of $\pm 75kHz$ at the highest modulating frequency $f_h = 15kHz$. This means that, we have a deviation ratio as determined in Eq. (4.30):

$$\delta = \frac{\Delta f_c}{f_h} = \frac{75x10^3}{15x10^3} = 5.0 \quad\quad\quad\quad\quad\quad\quad\quad (4.30)$$

4.3.2 Narrowband and broadband FM

When $m_f = 0.1$, it indicates that only one pair of significant sidebands appears. On the other side, when $m_f = 0.5$, it means that there are two pairs of significant sidebands. For small value of m_f, say $m_f < 1$, we have narrowband FM.

For values of $m_f \gg 1$, the sidebands cover a wide spectrum, but their amplitudes decrease. When the number of sidebands is greater than about 10, the number of significant sidebands does not depend on m_f.

In principle, the practical bandwidth, $(BW)_{practical}$, is around $2(\Delta f_c + f_h)$. That is, as given in Eq. (4.31):

OMAR FAKIH HAMAD

$$(BW)_{practical} \cong 2(\Delta f_c + f_h) \quad \text{..} \quad (4.31)$$

For $m_f = 5$; $\Delta f_c = 75kHz$; and $f_h = 15kHz$, we have a practical bandwidth, $(BW)_{practical}$, for a practical FM broadcast system, computed as:

$$(BW)_{practical} \cong 2(75x10^3 + 15x10^3)Hz$$
$$= 2x90xx10^3$$
$$= 180kHz$$

This shows the inherent wideband nature of FM. Nevertheless, the use of a large bandwidth welcomes an improved signal-to-noise-ratio (SNR). This constitutes one of the main advantages of FM compared to AM. The larger the value of Δf_c, the greater is the SNR improvement.

4.3.3 FM Generation and Detection

A direct method involves producing an FM wave by varying the frequency of the carrier by means of the modulating signal. It is done by using a non-linear device, say valve or transistor, to vary the capacitance across the tuned circuit of an oscillator. An indirect method invites the use of phase modulation.

4.3.3.1 Varactor modulator

It uses the variable capacitance of a p-n junction semiconductor which is referred as Varactor diode. A *dc* voltage is applied in the reverse direction across the p-n junction. The charges are drawn away from the junction leaving a depletion layer at the junction. This leads to a depleted part of charge carriers as shown in Fig. 4.7.

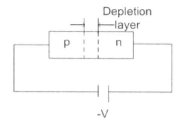

Fig. 4.7. Depletion layer

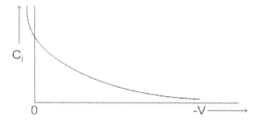

Fig. 4.8. C-V curve

There is a rise to a capacitance across the junction. This can be varied by varying the applied dc voltage. Therefore, at the junction, we will have a relation as given in Eq. (4.32): -

$$C_j \alpha \frac{1}{\sqrt{V}} \quad\text{---} \quad (4.32)$$

where V is the applied reverse voltage.

Usually, silicon diodes are characterized with C_j of between $150\,pF$ and $200\,pF$ with $1Volt$. C_j decreases to $50\,pF$ with $10Volts$ applied. The average variation ranges between $10\,pF$ and $15\,pF$ per $Volt$.

A typical Varactor modulator is given in Fig. 4.9.

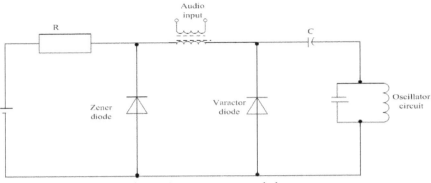

Fig. 4.9 Varactor modulator

Zener diode acts to stabilize the *dc* supply and it ensures that the mean oscillator frequency is not altered by supply voltage fluctuations. The Varactor capacitance, C_j, is varied by the modulating voltage, v_m, and the oscillator uses the feedback. The coupling capacitor provides dc isolation for the oscillator circuit.

4.3.4 FM transmitter

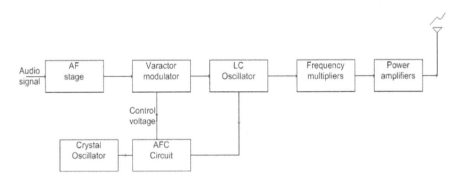

Fig. 4.10 FM transmitter employing a Varactor modulator

An FM transmitter employing a Varactor modulator is shown in Fig. 4.10. The audio signal is amplified in the AF stage and drives the Varactor modulator. The Varactor modulator varies the frequency of an LC oscillator, whose centre frequency is stabilised by crystal control via an AFC loop.

The initial carrier frequency is multiplied several times to bring it up into the VHF band by means of frequency multipliers. The output drives class-C RF power amplifiers with high efficiency, to give an FM output of a few kW.

Before going further, it is worth noting frequency ranges, classes, and bands which are used in daily applications of telecommunication Table 4.1 shows the frequency classes and their respective ranges while Table 4.2 show the frequency bands and their respective ranges

Table 4.1. Classes and ranges of frequencies

Class	Ranges
VLF (Very Low Frequency)	10kHz ~ 30kHz
LF (Low Frequency)	30kHz ~ 300kHz
MF (Medium Frequency)	300kHz ~ 3MHz
HF (High Frequency)	3MHz ~ 30MHz
VHF (Very High Frequency)	30MHz ~ 300MHz
UHF (Ultra High Frequency)	300MHz ~ 3GHz
SHF (Super High Frequency)	3GHz ~ 30GHz
EHF (Extremely High Frequency)	30GHz ~ 300GHz

It must be noted that, UHF and SHF are the prime classes that make microwave frequencies.

Table 4.2. Bands and ranges of frequencies

Class	Ranges
UHF (decimetre band)	0.3GHz ~ 1.0GHz
L (long band)	1.0GHz ~ 1.5GHz
S (short band)	1.5GHz ~ 3.9GHz
C (communication band)	3.9GHz ~ 8.0GHz
X (WW2 secret band)	8.0GHz ~ 12.5GHz
Ku (K-under or kay yoo or kutz)	12.5GHz ~ 18.0GHz

K (German "kurz" means short)	18.0GHz ~ 26.0GHz
Ka ("kurz-above")	26.0GHz ~ 40.0GHz
mm (millimetre wave, MMW or mmW)	40.0GHz ~ 300GHz

4.3.5 Direct FM and VCOs

Direct FM is straightforward and, hence, it requires nothing more than a voltage-controlled oscillator (VCO) whose oscillator frequency has a linear dependence on applied voltage. For $f_c \geq 1GHz$, that is, at microwave band, Klystron tube has linear VCO characteristics.

The principal advantage of direct FM is that we gain large frequency deviations without additional operations.

4.3.6 Phase modulators and indirect FM

We seldom transmit a PM wave, but we still like to generate wave using phase modulators. This is mainly because it is relatively easier to implement; the carrier can be supplied by a stable frequency source, and integrating the input signal to a phase modulator produces a frequency modulated output.

If we approximate $x_c(t)$ in Fig. 4.11 such that, in such away that $|\phi_\Delta x(t)| \ll 1 \ radian$, we have Eq. (3.33):

$$x_c(t) \cong A_c Cos\omega_c t - A_c \phi_\Delta x(t) Sin\omega_c t \ \text{.............................} \ (4.33)$$

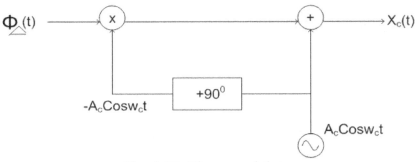

Fig. 4.11. Phase modulator

Large phase shift can be achieved by a switching circuit modulation as shown in Fig. 4.12:

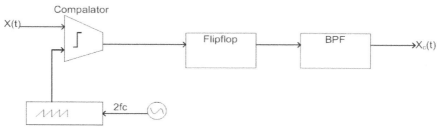

Fig. 4.12. Switching circuit modulator

4.3.7 Indirect FM transmitter

In Fig. 4.13, an integrator and a phase modulator constitute narrowband modulator with instantaneous frequency given in Eq. (4.34):

$$f_1(t) = f_{c1} + \frac{\phi_\Delta}{2\pi T} x(t) \quad\text{(4.34)}$$

where T is the integrator's proportionality constant.

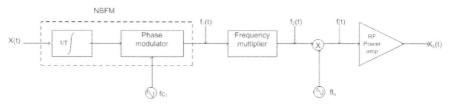

Fig. 4.13. Indirect FM Transmitter

The initial frequency deviation, f_Δ, is given by Eq. (4.35):

$$f_\Delta = \frac{\phi_\Delta}{2\pi T} \quad\text{(4.35)}$$

The desired value, f_Δ, is reached by a frequency multiplier. We can have n-fold multiplication of $f_1(t)$ to achieve $f_2(t)$. Therefore, we can have $f_2(t)$ can be expressed as in Eq. (4.36)

$$f_2(t) = nf_1(t) = nf_{c1} + f_\Delta x(t) \dots\dots\dots\dots\dots\dots (4.36)$$

Where

$$f_\Delta = n\left(\frac{\phi_\Delta}{2\pi T}\right) \dots\dots\dots\dots\dots\dots\dots\dots\dots (4.37)$$

The amount of multiplication required to get f_Δ leads to higher nf_{c1} than the desired carrier frequency. Therefore, a frequency converter is included. To translate the spectrum down to Eq. (4.38):

$$f_c = |nf_{c1} \pm f_{LO}| \dots\dots\dots\dots\dots\dots\dots\dots (4.38)$$

The final instantaneous frequency becomes as expressed in Eq. (4.39):

$$f(t) = f_c + f_\Delta x(t) \dots\dots\dots\dots\dots\dots\dots\dots (4.39)$$

4.3.8 Triangular wave FM

Triangular wave frequency modulation is in fact a modern and a novel method for FM. The technique overcomes the inherent problems of conventional VCOs and indirect FM systems. Triangular wave FM generates virtually distortion-less modulation at carrier frequencies up to 30MHz. It is well suited for instrumentation applications.

If we recall the fundamental expression as in Eq. (4.40):

$$x_c(t) = A_c Cos\,\theta_c(t) \dots\dots\dots\dots\dots\dots\dots (4.40)$$

where Eq. (4.41) holds:

$$\theta_c(t) = \omega_c t + \phi(t) - \phi(0) \dots\dots\dots\dots\dots\dots (4.41)$$

The condition in Eq. (4.41 is necessary as this guarantees that $\theta_c(0) = 0$.

Therefore, the instantaneous frequency, $f(t)$, is given by Eq. (4.42):

$$f(t) = \frac{1}{2\pi} \overset{o}{\theta_c}(t) = f_c + f_\Delta x(t) \dots \dots \dots \dots \dots (4.42)$$

Expressed in terms of $\theta_c(t)$, a unit-amplitude triangular FM signal can be expressed as in Eq. (4.43):

$$x_\Lambda(t) = \frac{2}{\pi} arcSin[Cos\,\theta_c(t)] \dots \dots \dots \dots \dots (4.43)$$

This is a triangular waveform when $\phi(t) = 0$ which can be mathematically appreciated with a simple observation.

Further, a periodic triangular function of θ_c is represented as in Eq. (4.44):

$$x_\Lambda(t) = \begin{cases} 1 - \dfrac{2}{\pi}\theta_c ; 0 < \theta_c < \pi \\ -3 + \dfrac{2}{\pi}\theta_c ; \pi < \theta_c < 2\pi \end{cases} \dots \dots \dots \dots (4.44)$$

Similarly, the derivation can be done for $\theta_c > 2\pi$.

Let us consider the modulation system in Fig. 4.14 to produce $x_\Lambda(t)$ from the voltage, $v(t)$, given by Eq. (4.45):

$$v(t) = \frac{2}{\pi} \overset{o}{\theta}(t) = 4[f_c + f_\Delta x(t)] \dots \dots \dots \dots \dots (4.45)$$

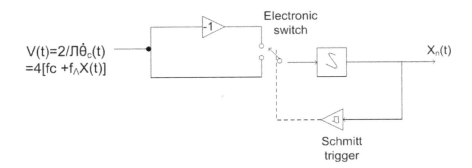

Fig. 4.14. Modulation system

The system essentially consists of an analog inverter, an integrator, and a Schmitt trigger to control an electronic switch.

Starting at $t = o$, with $x_\Lambda(0) = +1$ and the switch in the upper position. Then, for $0 < t < t_1$, we have $x_\Lambda(t)$ expressed as in Eq. (4.46) or Eq. (4.47):

$$x_\Lambda(t) = 1 - \int_0^t v(\lambda)d\lambda = 1 - \frac{2}{\pi}[\theta_c(t) - \theta_c(0)] \dots\dots\dots\dots \quad (4.46)$$

$$= 1 - \frac{2}{\pi}\theta_c(t); \ 0 < t < t_1 \dots\dots\dots\dots\dots\dots\dots\dots\dots \quad (4.47)$$

This implies that $x_\Lambda(t)$ traces out the downward ramp until t_1 when $x_\Lambda(t) = -1$ which implies that $\theta_c(t_1) = \pi$.

Now, the trigger throws the switch to the lower position and $x_\Lambda(t)$ can be expressed as in Eq. (4.48) or Eq. (4.49):

$$x_\Lambda(t) = -1 + \int_{t_1}^t v(\lambda)d\lambda = -1 + \frac{2}{\pi}[\theta_c(t) - \theta_c(t_1)] \dots\dots\dots \quad (4.48)$$

$$= -3 + \frac{2}{\pi}\theta_c(t); \ t_1 < t < t_2 \dots\dots\dots\dots\dots\dots\dots\dots\dots \quad (4.49)$$

The upward ramp, as in Fig. 4.15, continues until t_2 when $\theta_c(t_2) = 2\pi$ and $x_\wedge(t_2) = +1$

A sinusoidal FM wave is obtained from $x_\wedge(t)$ using a non-linear wave-shaper with transfer characteristics given by Eq. (4.50):

$$T[x_\wedge(t)] = A_c Sin\left[\left(\frac{\pi}{2}\right)x_\wedge(t)\right] \dotfill (4.50)$$

This does the inverse of Eq. (4.51):

$$x_\wedge(t) = \frac{2}{\pi} arcSin[Cos\,\theta_c(t)] \dotfill (4.51)$$

Otherwise, $x_\wedge(t)$ can be applied to a hard limiter to produce square wave FM in Fig. 4.16.

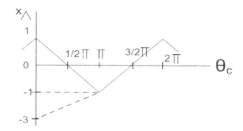

Fig. 4.15. The trace of $x_\wedge(t) = \dfrac{2}{\pi} arcSin[Cos\,\theta_c(t)]$

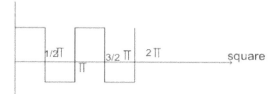

Fig. 4.16. Square wave FM

4.4 Frequency detection

A frequency detector is no different from a discriminator. It produces an output voltage that should vary linearly with the instantaneous frequency of the input. There are many different circuit designs for frequency detection.

Practically, there are four operational categories on which almost every circuit falls into. These are: (i) FM-to-AM conversion; (ii) Phase-shift discrimination; (iii) Zero-crossing detection; and (iv) Frequency feedback.

4.4.1 FM-to-AM conversion

Any device or circuit whose output is the time-derivative of the input produces FM-to-AM conversion.

Let us start with a function $x_c(t)$ given by Eq. (4.52):

$$x_c(t) = A_c Cos\theta_c(t) \text{-----------------------} (4.52)$$

where $\overset{o}{\theta_c}(t)$ is expressed by Eq. (4.53):

$$\overset{o}{\theta_c}(t) = 2\pi\left[f_c + f_\Delta x(t)\right] \text{-----------------------} (4.53)$$

This implies that, $\overset{o}{x_c}(t)$ is given by Eq. (4.54) or Eq. (4.55) as:

$$\overset{o}{x_c}(t) = -A_c \overset{o}{\theta_c}(t)Sin\theta_c(t) \text{-----------------------} (4.54)$$

$$= 2\pi A_c\left[f_c + f_\Delta x(t)\right]Sin\left[\theta_c(t) \pm 180^0\right] \text{-----------------------} (4.55)$$

Therefore, an envelope detector with an input $\overset{o}{x_c}(t)$ yields an output proportional to $f(t) = f_c + f_\Delta x(t)$.

A conceptual frequency detector, as described in Fig. 4.17, essentially includes a limiter (Lim) to remove any spurious amplitude variations from $x_c(t)$ and a *dc* block to remove the constant carrier-frequency effect from the output signal.

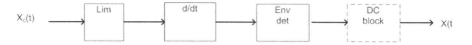

Fig. 4.17. Frequency detector with limiter and
FM-to-AM conversion

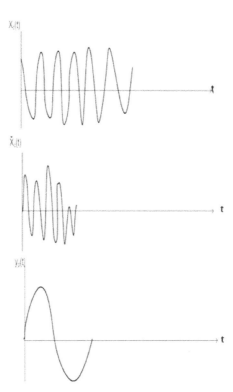

Fig. 4.18. Waveforms

The actual hardware implementation of FM-to-AM conversion considers the fact that an ideal differentiator has a transfer function, $H(f)$, obeying the expression in Eq. (4.56):

$$|H(f)| = 2\pi f \quad \text{..} \quad (4.56)$$

For this design, the related blocks that need to be considered include: (i) the slope detection block with a tuned circuit; (ii) the balanced discriminator circuit; and (iii) the frequency-to-voltage characteristic. The design involves circuits with linear phase response.

We start from the approximation for time differentiation given by Eq. (4.57):

$$\overset{o}{v}(t) \cong \frac{1}{t_1}[v(t) - v(t - t_1)] \quad \text{.......} \quad (4.57)$$

The situation to also consider is what if t_1 is small compared to the variation of $v(t)$.

From an FM wave in Eq. (4.58):

$$\overset{o}{\phi}(t) = 2\pi f_\Delta x(t) \quad \text{.......} \quad (4.58)$$

We get an expression that leads to Eq. (4.59):

$$\phi(t) - \phi(t - t_1) \cong t_1 \overset{o}{\phi}(t) = 2\pi f_\Delta t_1 x(t) \quad \text{.......} \quad (4.59)$$

The value of $\phi(t - t_1)$ can be obtained with the help of a delay line or, equivalently, a linear phase-shift network as diagrammatically elaborated in Fig. 4.19.

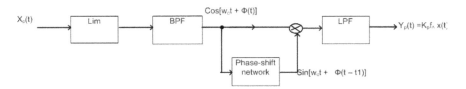

Fig. 4.19. Obtaining $\phi(t - t_1)$ with the help of a linear phase-shift network

The output is proportional to the expression in Eq. (4.60):

$$Sin[\phi(t) - \phi(t - t_1)] \cong \phi(t) - \phi(t - t_1) \quad \text{.......} \quad (4.60)$$

Assuming that t_1 is small enough at the extent that $|\phi(t) - \phi(t - t_1)| << \pi$, we can have an output of the LPF given by Eq. (4.61):

$$y_p(t) \cong K_p f_\Delta x(t) \dotfill \text{(4.61)}$$

4.4.2 Zero-crossing detector

Let us consider Figs. 4. 20, 21, and 22 that, respectively, show a Zero-crossing detector, the limiter output wave-forms, and the integrator's input wave-forms. The square wave FM signal from a hard limiter triggers a mono-stable pulse generator, which produces a short pulse of fixed amplitude A and duration τ at each upward or downward zero-crossing of the FM wave.

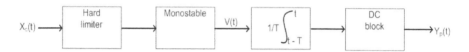

Fig. 4.20. Zero-crossing detector

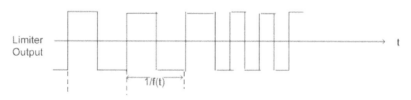

Fig. 4.21. Limiter output wave-forms

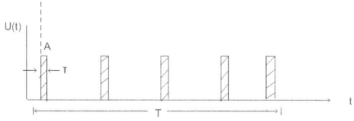

Fig. 4.22. Integrator's input wave-forms

If we consider a time interval T, such that $W \ll \dfrac{1}{T} \ll f(t)$, then the mono-stable output $v(t)$ looks like a rectangular pulse train with nearly constant period $\dfrac{1}{f(t)}$. Therefore, there will be $n_T \cong Tf(t)$ pulses in this interval. This means that we will have the validity of expression in Eq. (4.62):

$$\frac{1}{T}\int_{t-\tau}^{t} v(\lambda)d\lambda = \frac{1}{T}n_T A\tau \cong A\mathscr{f}(t) \dotfill (4.62)$$

This implies that expression in Eq. (4.63) will describe $y_p(t)$:

$$y_p(t) = K_p f_\Delta x(t) \dotfill (4.63)$$

To conclude, it is worth to mention two of the most widely used types of FM detectors. These are: (i) the Foster-Seeley circuit that gives better linearity, but it must be preceded by a limiter and (ii) the ratio detector that involves limiting and discrimination in the same circuit and it is largely used in domestic receivers.

4.5 Phase demodulation

As shown in Fig. 4.23, phase demodulation is done by means of an FM receiver (FM Rx) followed by an integrating network.

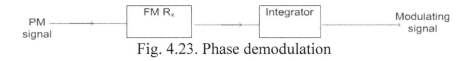

Fig. 4.23. Phase demodulation

Fig. 4.24 shows a feedback demodulation with FLL while Fig. 4.25 describes a feedback demodulation scheme with phase-locked loop (PLL) where the input signal is accepted at the phase-sensitive detector block. They are both use some kinds of feedback and thus the name *feedback demodulators*.

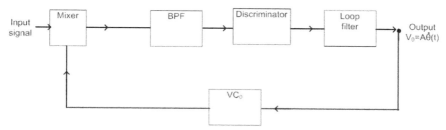

Fig. 4.24. Feedback demodulation with FLL

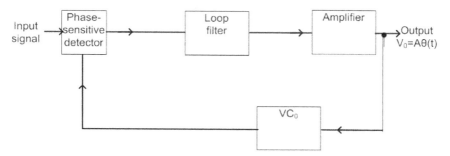

Fig. 4.25. Feedback demodulation with PLL

4.6 De-emphasis and pre-emphasis filtering

Technically, it has been proved that there is most severe FM interference at large values of $|f_i|$. Thus, there is a need for a method for improving system performance with selective *post-detection* filtering called *de-emphasis filtering*. Fig. 4.26 describes a schematic diagram of a complete FM demodulator with de-emphasis filtering.

Fig. 4.26. Complete FM demodulator with de-emphasis filtering

This will de-emphasize the higher frequency portion of the message band and reduce the more serious interference. Deemphasis filtering attenuates the higher frequency components of the message itself which causes distortion of the output signal. Therefore, corrective measures must be taken to compensate for de-emphasis distortion by *pre-distortion* or *pre-emphasizing* the modulation signal at the transmitter before modulation.

With $|f| \leq W$, the relationship in Eq. (4.64) must be satisfied for the two filters:

$$H_{pe}(f) = \frac{1}{H_{de}(f)} \quad \text{----------------------------} \quad (4.64)$$

This leads to a net undistorted transmission.

Pre-emphasis and de-emphasis filtering offers potential advantages whenever undesired contaminations tend to predominate at certain portions of the message band.

Let us take an example when HF are pre-emphasized in audio disk recording so that HF surface noise, say scratch, can be de-emphasized during playback. The FM de-emphasis filter is usually a simple first-order network having expression as in Eq. (4.65):

$$H_{de}(f) = \left[1 + j\left(\frac{f}{B_{de}}\right)\right]^{-1} \cong \begin{cases} 1; |f| << B_{de} \\ \dfrac{B_{de}}{jf}; |f| >> B_{de} \end{cases} \quad \text{--------------------} \quad (4.65)$$

In practice, the value of B_{de} is less than the message bandwidth, W.

At the transmitting end, the corresponding pre-emphasis filter function could be expressed as in Eq. (4.66):

$$H_{pe}(f) = \left[1 + j\left(\frac{f}{B_{de}}\right)\right] \cong \begin{cases} 1; |f| << B_{de} \\ \dfrac{jf}{B_{de}}; |f| >> B_{de} \end{cases} \quad \text{--------------------} \quad (4.66)$$

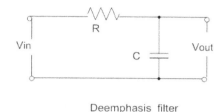

Deemphasis filter

Fig. 4.27. De-emphasis filter

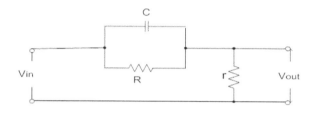

Preemphasis filter

Fig. 4.28. Pre-emphasis filter

Fig. 4.27 and Fig. 4.28, respectively, describe a de-emphasis filter and a pre-emphasis filter. It must be noted down that a pre-emphasis FM is a combination of FM and PM that leads to reduction of noise.

Chapter Five

Noise in Telecommunications Systems

5.1 Introduction

Noise is an unwanted, undesired signal which is not connected with the desired, wanted signal in any way. It is unpredictable in nature and, hence, we often talk about random noise signals. If it is predictable in nature, we consider that it is not random and it can be eliminated by proper design.

5.2 Sources of noise

Man made noise pick-up of undesired signals, the erratic natural disturbances which are irregularly occurring, and the fluctuation noise arising inside physical systems are the three main sources of noise in telecommunications systems.

The noise source in the first case may occur due to fault contacts, electrical appliances, ignition radiation, and fluorescent lighting. This can be eliminated by removing the source. For the case of noises arising from the second and the third sources, these are the non man-made types.

Noise from the second source may arise due to lightning, electric storms, or general atmospheric disturbances while the noise from the third source arises inside physical systems due to spontaneous fluctuations such as the thermal motion, characterised by Brownian movement, of free electrons inside a resistor, the random emission of electrons in vacuum tubes and random generation, recombination, and diffusion of electronic carriers—holes and electrons—in semiconductors.

By careful engineering, the effects of many unwanted signals can be reduced or eliminated completely. However, there always remain certain inescapable random signals that present a fundamental limit to system's performance. The unavoidable causes of electrical noise include the thermal motion of electrons in conducting media and the thermal noise which is always around and inside the telecommunications system.

5.2.1 Thermal noise

Thermal noise is the noise due to the random motion of charged particles in conducting media. This is what leads to *Johnson* noise or commonly known as *resistance noise*.

From Kinetic theory, the average energy of a particle at absolute temperature T is proportional to kT, where k is the Boltzmann constant.

When a metallic resistance R is at temperature T, random electron motion produces a noise voltage $v(t)$ at the open-circuited terminals. This voltage $v(t)$ has a Gaussian distribution with zero mean and variance with the relation expressed in Eq. (5.1):

$$\overline{v^2} = \sigma_v^{\,2} = \frac{2(\pi kT)}{3h} R \;\; Volts^2 \quad\text{--} \quad (5.1)$$

where

$k = Boltzmann\ constant = 1.37x10^{-23}J/deg$
$h = Planck's\ constant = 6.62\ x10^{-34}J.s$
$T = Temperature,\ measured\ in\ Kelvins.$

The spectral density of thermal noise is given by Eq. (5.2) and diagrammatically represented as in Fig. 5.1.

$$G_v(f) = \frac{2Rh|f|}{(e^{h|f|/kT} - 1)} \;\; Volts^2\,/\,Hz \quad\text{------------------------------------} \quad (5.2)$$

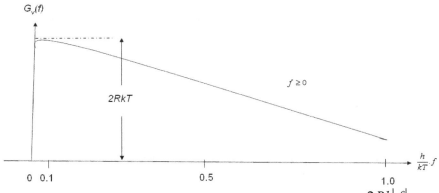

Fig. 5.1. Spectral density of thermal noise, $G_v(f) = \dfrac{2Rh|f|}{(e^{h|f|/kT} - 1)}$

At *low* frequencies, with the help of series approximation, we can have Eq. (5.3);

$$G_v(f) \cong 2RkT\left(1 - \frac{h|f|}{2kT}\right) \quad\text{-----------------} \quad (5.3)$$

where $|f| << \dfrac{kT}{h}$.

The approximations in Eq. (5.3) are hardly needed by telecommunications engineers. So, let the room temperature or the standard temperature be as in a relation expressed in Eq. (5.4):

$$T_o \underline{\Delta} 290K = (63^o F) \quad\text{-----------------} \quad (5.4)$$

This means that, we will have $kT_o \cong 4x10^{-21}W.s$

If the resistance is at T_o, then $G_v(f)$ will be constant for $|f| < 0.1\dfrac{kT_o}{h} \cong 10^{12}Hz$. This is a part of infrared portion of the EM spectrum. It is very far from the point where conventional electrical components have ceased to respond.

Therefore, Eq. (5.5) expresses the spectral density of thermal noise when it is constant:

$$G_v(f) \cong 2RkT \; Volts^2 \, / \, Hz \quad \text{---} \quad (5.5)$$

5.2.2 White noise and filtered noise

Besides thermal resistance, many other types of noise sources are Gaussian and have a flat spectral density over a wide range of frequencies. When we have all frequency components in equal proportion, we have *white noise*. The spectral density of white noise in general is given as in Eq. (5.6):

$$G(f) = \frac{\eta}{2} \quad \text{--} \quad (5.6)$$

which signifies half of the power at the positive frequency and the half at the negative frequency with η standing as the positive-frequency power density. Fig. 5.2 shows the diagrammatical presentation of the power spectrum of the said white noise.

The auto-correlation function is given as in Eq. (5.7) and Fig. 5.3 shows that the autocorrelation of white noise at the origin $(R(\tau \neq 0) = 0)$

$$R(\tau) = \int_{-\infty}^{+\infty} \frac{\eta}{2} e^{j\omega\tau} df = \frac{\eta}{2} \delta(\tau) \quad \text{----------------------------} \quad (5.7)$$

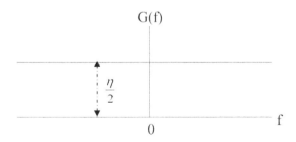

Fig. 5.2. Power spectrum of white noise

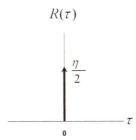

Fig. 5.3. Autocorrelation of white noise ($R(\tau \neq 0) = 0$)

The power density η depends on two factors—the type of noise source, and the type of spectral density. If the source is a thermal resistor, then we have, respectively, η_v, η_i, and η_a for the mean square voltage, the mean square current, and the available power expressed as in Eqs. (5.8), (5.9), and (5.10):

$$\eta_v = 4RkT \quad\quad\quad (5.8)$$

$$\eta_i = \frac{4kT}{R} \quad\quad\quad (5.9)$$

$$\eta_a = kT \quad\quad\quad (5.10)$$

By definition, any thermal noise source has $\eta_a = kT$. Other white noise sources are non-thermal, as the η_a is unrelated to a physical temperature. The noise temperature of any white noise source, thermal or non-thermal, is defined as in Eq. (5.11):

$$T_N \triangleq \frac{2G_a(f)}{k} = \frac{\eta_a}{k} \quad\quad\quad (5.11)$$

where η_a is the maximum noise power the source can deliver per unit frequency and it is given by Eq. (5.12):

$$\eta_a = kT_N \quad\quad\quad (5.12)$$

where T_N is not necessarily a physical temperature.

For better understanding, let us consider an example of certain electronic noise generation systems that have $T_N = 10T_o \cong 3000K \cong 5000^0 F$, but the devices are not that hot. If we consider Gaussian white noise $x(t)$ with spectral density $G_x(f) = \dfrac{\eta}{2}$ applied to a linear time-invariant (LTI) filter having transfer function $H(f)$. Then, the resulting output, $y(t)$, will be Gaussian noise described by Eq. (5.13):

$$G_y(f) = \frac{\eta}{2}|H(f)|^2 \quad\quad\quad\quad\quad\quad (5.13)$$

Mathematics shows that the spectral density of filtered white noise takes the shape of $|H(f)|^2$. Filtering white noise produces coloured noise with frequency content primarily in the range passed by the filter.

The autocorrelation $R_y(\tau)$ is given by the expression in Eq. (5.14):

$$R_y(\tau) = \frac{\eta}{2}\mathfrak{I}^{-1}\left[|H(f)|^2\right] \quad\quad\quad\quad\quad (5.14)$$

The output power $\overline{y^2}$ is directly proportional to the filter's band-width as expressed in Eq. (5.15):

$$\overline{y^2} = \frac{\eta}{2}\int_{-\infty}^{\infty}|H(f)|^2\,df = \frac{\eta}{2}[f]_{-B}^{+B} \quad\quad\quad\quad (5.15)$$

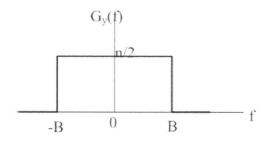

Fig. 5.4. The trace of $G_y(f)$ vs. f

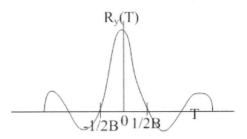

Fig. 5.5. The trace of $R_y(\tau)$ vs. τ

Fig. 5.4 shows the trace of $G_y(f)$ against f while Fig. 5.5 represents the trace of $R_y(\tau)$ against τ. If $H(f)$ is an ideal low pass function and we have a unit gain with bandwidth B, then, $G_y(f)$ and $R_y(\tau)$ are expressed as in Eq. (5.16) and (5.17), respectively:

$$G_y(f) = \frac{\eta}{2}\pi\left(\frac{f}{2B}\right) \quad\text{------------------------------------(5.16)}$$

$$R_y(\tau) = \eta B Sinc(2B\tau) \quad\text{------------------------------(5.17)}$$

5.3 Probability density of noise

Statistically, if the noise $n(t)$ has a power spectral density $\frac{\eta}{2}$ over a bandwidth B, then, Eq. (5.18) is valid:

$$\sigma^2 = \eta B \quad\text{--(5.18)}$$

The probability density at any fixed time, $f(n)$, is given by Eq. (5.19):

$$f(n) = \frac{1}{\sqrt{2\pi\eta B}}e^{-n^2/2\eta B} \quad\text{------------------------(5.19)}$$

5.4 Noise in AM systems

Consider Fig. 5.6 which represents an existence of noise in AM system with the outputs of the baseband filter being real signal, S_o, and the noise signal, N_o.

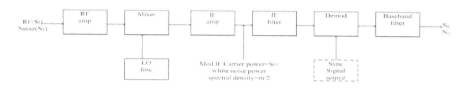

Fig. 5.6. Noise in AM system

If we consider that the upper-side band of SSB-SC system is being used.

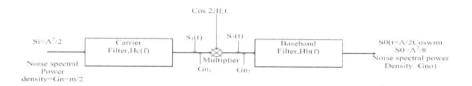

Fig. 5.7. Noise spectral power density

Fig. 5.7 shows a diagrammatical representation of the system for computation of noise spectral power density. From the figure, the received signal can be expressed as in Eq. (5.20):

$$S_i(t) = ACos[2\pi(f_c + f_m)t]$$ .... (5.20)

The output of the multiplier is given by Eq. (5.21):

$$S_2(t) = S_1(t)Cos\omega_c t = \frac{A}{2}Cos[2\pi(2f_c + f_m)t] + \frac{A}{2}Cos2\pi f_m t \quad (5.21)$$

However, the baseband filter will allow only the difference-frequency terms. Therefore, the output signal will be represented as in Eq. (5.22):

$$S_o(t) = \frac{A}{2}Cos2\pi f_m t \text{------------------------------ (5.22)}$$

The expression in Eq. (5.22) is a modulating signal amplified by half.

The input signal power is given by Eq. (5.23):

$$S_i = \frac{A^2}{2} \text{--- (5.23)}$$

The output signal power is, therefore, expressed by Eq. (5.24):

$$S_o = \frac{1}{2}\left(\frac{A}{2}\right)^2 = \frac{A^2}{8} \text{------------------------ (5.24)}$$

Therefore, the ratio $\frac{S_o}{S_i}$ becomes $\frac{S_o}{S_i} = \frac{A^2/8}{A^2/2} = \frac{1}{4}$

The result of $\frac{S_o}{S_i}$ is entirely general.

S_i and S_o are properly the total power, independently of whether a single or many spectral components are involved.

5.4.1 Noise power

Let that the input noise is white and of spectral density $\frac{\eta}{2}$. If G_{no} is the noise transmitted by the baseband filter and if we set G_{n1} as the noise input to the multiplier such that G_{n2} can be expressed in terms of G_{n1} as in Eq. (5.25):

$$G_{n2} = G_{n1}xCos2\pi f_c t \text{------------------------------ (5.25)}$$

Therefore, output noise, N_o, can be expressed as in Eq. (5.26)

$$N_o = 2f_M \frac{\eta}{8} = \frac{\eta f_M}{4} \quad\quad\quad (5.26)$$

5.4.2 Signal-to-noise ratio (SNR)

For the case of SSB-SC, sSignal-to-noise ratio (SNR) at the output can be calculated as in Eq. (5.27):

$$\frac{S_o}{N_o} = \frac{S_i/4}{\eta f_M/4} = \frac{S_i}{\eta f_M} \quad\quad\quad (5.27)$$

SNR which is simply the ratio $\dfrac{S_o}{N_o}$ serves as a figure of merit of the performance of a telecommunication system. As an SNR increases, it becomes easier to distinguish and to reproduce the modulating signal without error or confusion.

For the case of DSB-SC, the input signal is given by Eq. (5.28):

$$S_i = \frac{A^2}{2} \quad\quad\quad (5.28)$$

and the output signal is expressed as in Eq. (5.29)

$$S_o = \frac{A^2}{4} = \frac{S_i}{2} \qu\quad\quad (5.29)$$

Therefore, the signal-to-noise ratio, *SNR*, is given by considering Eq. (5.30):

$$\frac{S_o}{N_o} = \frac{S_i/2}{N_o} \qu\quad\quad (5.30)$$

However,

$$N_o = \frac{\eta}{4}(2f_M) = \frac{\eta f_M}{2} \quad\quad\quad (5.31)$$

Therefore, the expression for *SNR* is given by Eq. (5.32):

$$SNR = \frac{S_o}{N_o} = \frac{S_i/2}{\eta f_M/2} = \frac{S_i}{\eta f_M} \quad \text{(5.32)}$$

The important observation is, therefore, that the *SNR* for SSB-SC is exactly as for DSB-SC. It is left as homework to analyze the figure of merit, *SNR*, for the case of DSB-with carrier.

5.5 Noise in FM systems

Consider a limiter-discriminator used to demodulate an FM signal as shown in Fig. 5.8.

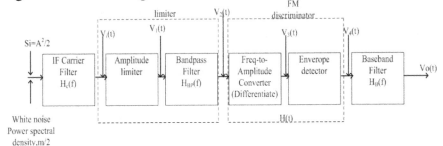

Fig. 5.8. Limiter-discriminator demodulator

To find the output signal and noise power, let the input signal to the *IF* carrier filter be given as in Eq. (5.33):

$$S_i(t) = ACos\left[\omega_c t + k \int_{-\infty}^{t} m(\lambda)d\lambda \right] \quad \text{(5.33)}$$

such that $m(t)$ is the frequency-modulating baseband waveform.

It is assumed that the signal is embedded in *additive white Gaussian noise* of power spectral density $\frac{\eta}{2}$. The Carson's rule for bandwidth requires that a bandwidth, B, be given as in Eq. (5.34):

$$B = 2\Delta f + 2 f_M \quad \text{(5.34)}$$

The filter passes the signal with negligible distortion and eliminates all noise outside the bandwidth B. The signal plus its accompanying noise is ideally limited, discriminated and appearing at the output as a signal $S_o(t)$ and a noise waveform $n_o(t)$.

It can be shown that, when the SNR is high, the noise does not affect the output-signal power. The noise can be ignored in calculating the output-signal power.

At the output of the limiter, the signal $S_2(t)$ is given by Eq. (5.35):

$$S_2(t) = A_L Cos\left[\omega_c t + k \int_{-\infty}^{t} m(\lambda)d\lambda \right] \quad (5.35)$$

If we set

$$\phi(t) = k \int_{-\infty}^{t} m(\lambda)d\lambda \quad (5.36)$$

At the discriminator's output, we find $S_4(t)$ expressed as in Eq. (5.37):

$$S_4(t) = \alpha\omega_c + \alpha\frac{d}{dt}\phi(t) \quad (5.37)$$

This implies that:

$$S_4(t) = \alpha\omega_c + \alpha k m(t) \quad (5.38)$$

Since the baseband filter rejects the *dc* components and passes the signal components without distortion, the output signal is given in Eq. (5.39):

$$S_o(t) = \alpha k m(t) \quad (5.39)$$

The output-signal power is given in Eq. (5.40):

$$S_o(t) = \alpha^2 k^2 \overline{m^2(t)} \quad (5.40)$$

5.5.1 Output noise power

For $|f| \le \dfrac{B}{2}$, the output spectral density of noise $n_4(t)$ is given by Eq. (5.41):

$$G_{n4}(f) = \frac{\alpha^2 \omega^2}{A^2} \eta \hspace{2cm} (5.41)$$

The output noise power N_o is, hence, given by expression in Eq. (4.42):

$$N_o = \int_{-f_M}^{f_M} G_{n4}(f)\,df \hspace{2cm} (5.42)$$

The graphical representation of $G_{n4}(f)$ vs. f can easily be drawn and visualized for the interval:

$$-\frac{B}{2} \le f \le \frac{B}{2} \hspace{2cm} (5.43)$$

Therefore,

$$N_o = \frac{\alpha^2 \eta}{A^2} \int_{-f_M}^{f_M} 4\pi^2 f^2\,df = \frac{\alpha^2 \eta}{A^2}\left[\frac{4\pi^2 f^3}{3}\right]_{-f_M}^{f_M} \hspace{1cm} (5.44)$$

$$= \frac{\alpha^2 \eta}{A^2}\left[\frac{4\pi^2 f_M^3}{3} + \frac{4\pi^2 f_M^3}{3}\right]$$

This results into,

$$N_o = \frac{8\pi^2 \alpha^2 \eta}{3A^2} f_M^3 \hspace{2cm} (5.45)$$

5.5.2 Output signal-to-noise ratio

From the expression for S_o and N_o, we can compute the output SNR, $\dfrac{S_o}{N_o}$, as in Eq. (4.46):

$$\frac{S_o}{N_o} = \frac{\alpha^2 k^2 \overline{m^2(t)}}{\dfrac{8\pi^2 \alpha^2 \eta}{3A^2} f_M^{\,3}} = \frac{3}{4\pi^2} \cdot \frac{k^2 \overline{m^2(t)}}{f_M^{\,2}} \cdot \frac{A^2/2}{\eta f_M} \quad\text{(5.46)}$$

Consider the case that the modulating signal $m(t)$ is sinusoidal and produces a frequency deviation Δf, the input signal $S_i(t)$ is given by Eq. (5.47):

$$S_i(t) = ACos(\omega_c t + \frac{\Delta f}{f_m} Sin2\pi f_m t) \quad\text{(5.47)}$$

This implies that:

$$km(t) = 2\pi\Delta f Cos 2\pi f_m t \quad\text{(5.48)}$$

Squaring the two sides of the equation above, we get Eq. (5.49):

$$k^2 \overline{m^2(t)} = \frac{4\pi^2 (\Delta f)^2}{2} = 2\pi^2 (\Delta f)^2 \quad\text{(5.49)}$$

Therefore, the SNR is obtained as in Eq. (5.50):

$$\frac{S_o}{N_o} = \frac{3}{2}\left(\frac{\Delta f}{f_M}\right)^2 \cdot \frac{A^2/2}{\eta f_M} = \frac{3}{2} \cdot \beta^2 \cdot \frac{S_i}{N_M} \quad\text{(5.50)}$$

where $N_M = \eta f_M$ is the noise power at the input in the baseband bandwidth f_M.

5.6 Comparison of angular and linear modulation systems

Based on the systems applications and the performance required, it might be helpful to systematically compare between the angular and linear modulation systems. For the purpose, the *figure of merit*, γ, is introduced.

For the case of AM, it is mathematically defined and expressed as in Eq. (5.51):

$$\gamma_{AM} \triangleq \frac{\left(S_o/N_o\right)}{\left(S_i/N_M\right)} \quad\text{--} \quad (5.51)$$

For the case of AM systems

  (a) $\gamma_{AM} = 1$ for the case of SSB-SC,

  (b) $\gamma_{AM} = 1$ for the case of DSB-SC,

  (c) $\gamma_{AM} = \dfrac{\overline{m^2(t)}}{1 + \overline{m^2(t)}}$ for the case of DSB, and

  (d) $\gamma_{AM} = \dfrac{m^2}{2 + m^2}$ for the case of DSB with sinusoidal modulation.

For the case of FM systems, the figure of merit, γ_{FM}, is expressed as in Eq. (5.52):

$$\gamma_{FM} \triangleq \frac{\left(S_o/N_o\right)}{\left(S_i/N_M\right)} = \frac{3}{2}\beta^2 \quad\text{------------------------------} \quad (5.52)$$

For the case of DSB with carrier, which is of much interest, if a 100% modulation of sinusoidal waveform is considered, then γ_{AM} and γ_{FM} are given as in Eq. (5.53) and Eq. (5.54), respectively:

$$\gamma_{AM} = \frac{1^2}{2+1^2} = \frac{1}{3} \quad\text{--}\quad (5.53)$$

and

$$\gamma_{FM} = \frac{3}{2}\beta^2 \quad\text{--}\quad (5.54)$$

Therefore, we have a comparison ratio for the figure of merits of FM and AM given by Eq. (5.55):

$$\frac{\gamma_{FM}}{\gamma_{AM}} = \frac{9}{2}\beta^2 \quad\text{-------------------------------------}\quad (5.55)$$

It must be noted that assumptions have been made that there is equal input-noise power spectral density, $\frac{\eta}{2}$, with equal baseband bandwidth, f_M, and equal input-signal power, S_i.

The comparison on the basis of equal signal power measured when the modulation $m(t) = 0$, is given by Eq. (5.56):

$$\frac{\gamma_{FM}}{\gamma_{AM}} = 3\beta^2 \quad\text{--------------------------------------}\quad (5.56)$$

The above observations convince that FM offers the possibility of improved *SNR* over AM. The improvement begins when $\frac{9}{2}\beta^2 \cong 1$. That is, when $\beta \cong 0.5$. Or when $3\beta^2 \cong 1$. That is, when $\beta \cong 0.6$. As β increases, the improvement becomes more pronounced, but greater bandwidth is required.

As an example, let β is large enough, so that Eq. (5.57) is reasonable:

$$B_{FM} = 2(\beta+1)f_M \quad\text{------------------------------------}\quad (5.57)$$

That means, the above expression can be approximated to: $B_{FM} \cong 2\beta f_M$ and $B_{AM} \cong 2f_M$

Hence, we will have expression for β as in Eq. (5.58)

$$\beta = \frac{B_{FM}}{2f_M} = \frac{B_{FM}}{B_{AM}} \text{---} (5.58)$$

Therefore, Eq. (5.59) follows:

$$\frac{\gamma_{FM}}{\gamma_{AM}} = \frac{9}{2}\left(\frac{B_{FM}}{B_{AM}}\right)^2 \text{----------------------------} (5.59)$$

The expression in Eq. (5.59) concludes that each increase in band-width by a factor of 2 increases $\frac{\gamma_{FM}}{\gamma_{AM}}$ by a factor of 4, that is, 6dB.

5.6.1 Noise figure

Before starting to analyze the concept of noise figure, let us put some valid assumptions. It is firstly assumed that the noise at the input to a two-port may be represented as being due to a resistor at the two-port input. Secondly, we assume that the resistor is at room temperature, $T_o = 290^o K$. Finally, we assume that the two-port itself is entirely noiseless.

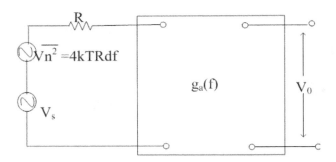

Fig. 5.9. Noise at the input to a two-port

Let us consider Fig. 5.9 that shows noise at the input to a two-port. The available noise-power spectral density will be given by Eq. (5.60):

$$G'_{ao} = g_a(f)\left(\frac{kT_o}{2}\right) \quad\quad (5.60)$$

Nevertheless, the actual noise-power spectral density is $G_{ao} > G'_{ao}$. The ratio $\frac{G_{ao}}{G'_{ao}} = F$ is the noise figure of the two-port.

Thus, we can define F as in expression in Eq. (5.61):

$$F \triangleq \frac{G_{ao}}{G'_{ao}} = \frac{G_{ao}}{g_a(f)\left(\frac{kT_o}{2}\right)} \quad\quad (5.61)$$

For noiseless two-port, we have $F = 1$ which means $0dB$. Otherwise, $F > 1$

Then, G_{ao} is given by Eq. (5.62):

$$G_{ao} = g_a(f)\frac{k}{2}(T + T_e) \quad\quad (5.62)$$

where T_e is the effective-input-noise temperature.

From $G_{ao} = g_a(f)\frac{k}{2}(T + T_e)$, let $T = T_o$, then, T_e and F are related as in Eq. (5.63):

$$T_e = T_o(F - 1) \quad\quad (5.63)$$

This means that what we call spot noise figure, F, is given by Eq. (5.64):

$$F = 1 + \frac{T_e}{T_o} = \frac{T_e + T_o}{T_o} \quad\quad (5.64)$$

The average noise figure over a frequency range from f_1 to f_2 is given by Eq. (5.65):

$$\overline{F} = \frac{\int_{f_1}^{f_2} g_a(f)F(f)df}{\int_{f_1}^{f_2} g_a(f)df} \quad\text{------------------------------ (5.65)}$$

For uniform power spectral density in a frequency range from f_1 to f_2, $\overline{F}$ is given b Eq. (5.66):

$$\overline{F} = \frac{S_i/N_i}{S_o/N_o} \quad\text{------------------------------ (5.66)}$$

If $S_o = g_a S_i$, then, if $F = \overline{F}$, the Eq. (5.67) is true:

$$F = \frac{S_i/N_i}{g_a S_i/N_o} = \frac{S_i}{N_i}\frac{N_o}{g_a S_i} = \frac{1}{g_a}\frac{N_o}{N_i} \quad\text{------------------------------ (5.67)}$$

But $N_o = g_a N_i + N_{tp}$ where $g_a N_i$ corresponds to the noise presence at the input and N_{tp} is due to the two-port. Therefore, F is given by Eq. (5.68):

$$F = 1 + \frac{N_{tp}}{g_a N_i} \quad\text{------------------------------ (5.68)}$$

This gives Eq. (5.69);

$$N_{tp} = g_a(F-1)N_i \quad\text{------------------------------ (5.69)}$$

5.6.1.1 Noise figure and equivalent noise temperature of a cascade

Consider a cascade of 2 two-ports, as in Fig. 5.10, with a noise source at the input of noise temperature T_o.

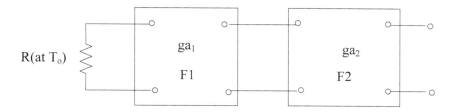

Fig. 5.10. **A** cascade of 2 two-ports with a noise source at the input of noise temperature T_o

The noise output of the first stage due to the source N_i is:

$$g_{a1}g_{a2}N_i \quad\text{(5.70)}$$

While the noise output of the first stage due to the noise generated within this first two-port is:

$$g_{a1}(F_1-1)N_i \quad\text{(5.71)}$$

Noise at the output of the second stage is:

$$g_{a1}g_{a2}(F_1-1)N_i \quad\text{(5.72)}$$

The noise output due to the noise generated within the second two-port is:

$$g_{a2}(F_{21}-1)N_i \quad\text{(5.73)}$$

The total noise output is, therefore, expressed as in Eq. (5.74):

$$N_o = g_{a1}g_{a2}N_i + g_{a1}g_{a2}(F_1-1)N_i + g_{a2}(F_2-1)N_i \quad\text{(5.74)}$$

The overall noise figure of the cascade is given in Eq. (5.75):

$$F = \frac{1}{g_a}\frac{N_o}{N_i} = \frac{1}{g_{a1}g_{a2}}\frac{N_o}{N_i} = F_1 + \frac{F_2-1}{g_{a1}} \quad\text{(5.75)}$$

For a cascade of k stages, F is given as in Eq. (5.76):

$$F = F_1 + \frac{F_2 - 1}{g_{a1}} + \frac{F_3 - 1}{g_{a1}g_{a2}} + \ldots + \frac{F_k - 1}{g_{a1}g_{a2}\cdots g_{a(k-1)}} \ldots\ldots\ldots\ldots\ldots (5.76)$$

The equivalent temperature of the cascade, T_e, is related to the equivalent temperatures and available gains of the individual stages by the expression in Eq. (5.77):

$$T_e = T_{e1} + \frac{T_{e2}}{g_{a1}} + \frac{T_{e3}}{g_{a1}g_{a2}} + \ldots + \frac{T_{ek}}{g_{a1}g_{a2}\cdots g_{a(k-1)}} \ldots\ldots\ldots\ldots\ldots (5.77)$$

5.7 Circuit noise and circuit noise level

5.7.1 Circuit noise

Considering the fact that a telecommunications system is a synergy of many separable segments of electronics and electric circuits, circuit noise is abstractly defined as the noise generated within these electronics and electric circuits and the noise may come from many sources involving several different physical phenomena. In general, circuit noise reflects the device noise found in common systems circuit, characterized by simple or complex resolvable circuit models.

5.7.2 Circuit noise level

If we can devise mechanisms that can separately provide a circuit noise at any point in a transmission media of telecommunications system and a noise at a nominal reference point, then a circuit noise level is the ratio of the circuit noise at any point in a transmission media of telecommunications system to a nominal reference point. This ratio is either expressed in decibels, above the reference noise, expressed in *dbrn*, or in adjusted decibels, expressed in *dba*. The quantity in *dba* reflects a specified adjustment due to external interference. In short, considering any point in a transmission system, the ratio of the circuit noise at that point to an arbitrary level chosen as a reference is what we refer as circuit noise level.

Conventionally, the practical circuit noise level is usually expressed in *dBrn0*, signifying the reading of a circuit noise meter, or in *dBa0*, signifying circuit noise meter reading adjusted to represent an interfering effect under specified conditions. Some analog circuit components, like amplifiers and converters are the main sources of the interferences which cause noise in a circuit. We identify the lowest signal level that a circuit can reliably process by calculating the circuit noise during the design stage.

Noise identification process can be a complex task because it must involve derivation and solving of many simple and sometimes complex equations written after identifying and accounting for uncorrelated and correlated noise sources in a given circuit system. The best and possible practice is to assume that all of the circuit's noise sources are correctly identified and are configured for optimum noise performance.

Chapter Six

Multiplexing

6.1 Introduction

Multiplexing is the sending of a number of separate signals together, over the same cable, channel, link, or bearer, simultaneously and without interference. There are two broad kinds of multiplexing. These are (i) time-division multiplexing (TDM) and (ii) frequency-division multiplexing (FDM).

Time-division multiplexing (TDM) is a method of interleaving, in the time-domain, pulses belonging to different transmissions. The advantage is taken of the fact that pulses are generally narrow, and separation between successive pulses is rather wide. If the two ends of the link are *synchronized,* the wide spaces can be used for pulses belonging to other transmissions.

Frequency-division multiplexing (FDM) combines continuous or analog signals. A sum of 12 or 16 channels are combined into a group, 5 groups into a subgroup, and so on, using frequencies and arrangements that are standard on a worldwide scale. Each group, super-group or larger aggregate is then sent as a whole unit on one microwave link, cable or other broadband systems.

6.2 Frequency-division multiplexing (FDM)

When more than one channels are needed between the same two points, economy of transmission is to be planned. A technique to end all the messages on one transmission facility is needed. The principle of FDM is illustrated in Fig. 6.1 where three signals are multiplexed.

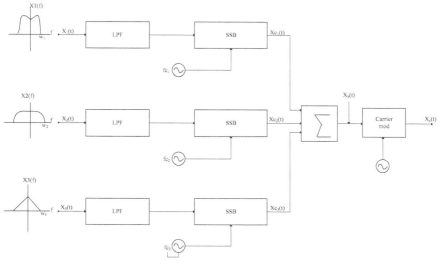

Fig. 6.1. FDM of $X_1(f)$, $X_2(f)$, and $X_3(f)$

Fig. 6.2.shows guard bands in FDM of $X_1(f)$, $X_2(f)$, and $X_3(f)$.

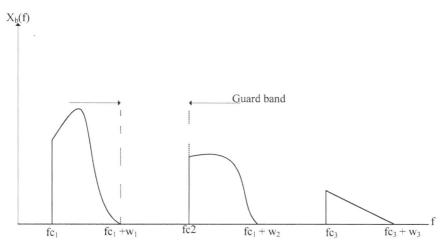

Fig. 6.2.Guard bands in FDM of $X_1(f)$, $X_2(f)$, and $X_3(f)$

Several input messages individually modulate the subscribers f_{c1}, f_{c2}, f_{c3}, and so on, after passing through LPFs to limit the message bandwidths. Sub-carrier modulation is often SSB, but any of the CW modulation techniques could be employed or a mixture of them. The modulated signals are then summed to produce the baseband signal, with spectrum $X_b(f)$.

The multiplexing operation has assigned a slot in the frequency domain for each of the individual messages in modulated form. The baseband signal may then be transmitted directly or used to modulate a transmitted carrier of frequency f_c.

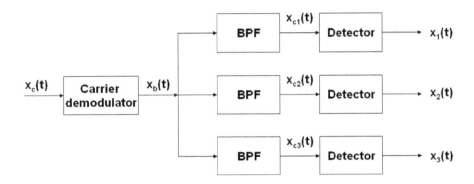

Fig. 6.3.Demodulation of FDM

The demodulation of FDM is accomplished in 3 steps as depicted in Fig. 6.3. The carrier demodulator reproduces the baseband signal $X_b(f)$; the demodulated sub-carriers are separated by a bank of bandpass filters in parallel; and the messages are then individually detected.

The major practical problem of FDM is cross-talk, the unwanted coupling of one message into another. This is due to non-linearities in the system.

A practical case of multiplexing is the AT&T type L4 carrier system which has 3600 voice channels of $W = 4kHz$ each multiplexed together for transmission via coaxial cable with SSB—both USSB and LSSB—being used. The final baseband spectrum corresponding to $0.5 \sim 17.5MHz$ including pilot carrier and guard bands. All sub-carriers are multiples of $4kHz$ with $512kHz$ pilot providing synchronization.

As a matter of practical consideration, it must be noted that an excessive guard-band width might be needed. To avoid this, the multiplexing is done by groups in four stages as roughly shown in Fig. 6.4.

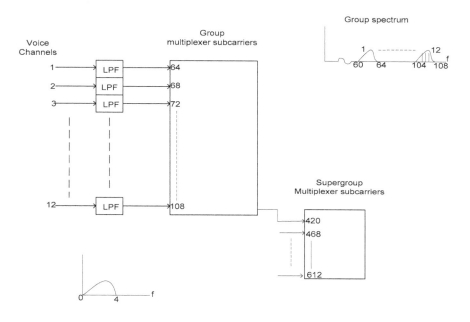

Fig. 6.4. FDM in 4 stages

The first two stages of telephone multiplexing includes the following operations: (i) 12 input voice channels are multiplexed resulting into a group of $48kHz$ at $60kHz$; (ii) Five such groups are multiplexed resulting into a super-group from $312 \sim 552kHz$, and so on.

Table 6.1 shows the AT&T FDM hierarchy where groups, super-groups, master-groups, and jumbo-groups have been mentioned against their frequency ranges, bandwidths, and number of voice channels

Table 6.1. AT&T FDM hierarchy

Designation	Frequency range (L4 system)	Bandwidth	Number of voice channels
Group	$60 \sim 108 kHz$	$48 kHz$	12
Super-group	$312 \sim 552 kHz$	$240 kHz$	60
Master-group	$564 \sim 3084 kHz$	$2.52 MHz$	600
Jumbo-group	$0.5 \sim 17.5 MHz$	$17 MHz$	3600

6.3 Time-division multiplexing (FDM)

While sampling the waveform, there is time when the waveform is off and, hence, time is left for other purposes. Sample values from several different signals can be interlaced into a single waveform. This is the principle of TDM.

6.3.1 TDM systems

The essential features of TDM are shown in the block diagram in Fig. 6.5:

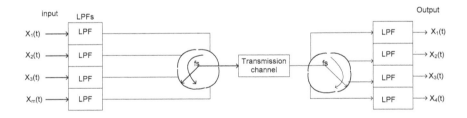

Fig. 6.5. TDM of $x_1(t)$, $x_2(t)$, and $x_3(t)$

Several input signals are pre-filtered by the bank of input LPFs and sampled *sequentially*. The rotating sampling switch, *commutator*, at the transmitter extracts one sample from each input per revolution. The output is a PAM waveform that contains the individual samples periodically interlaced in time.

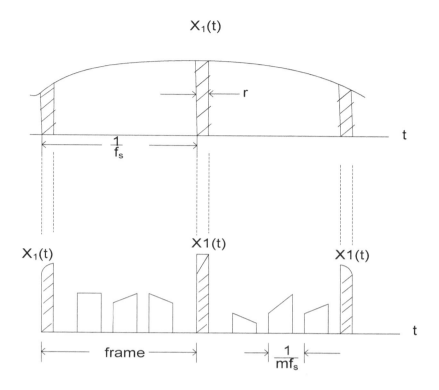

Fig. 6.6. Multiplexed PAM waves

At the receiver, there is a similar rotary switch, *de-commutator,* or *distributor* that separates the samples and distributes them to another bank of LPFs for the individual messages reconstruction.

If all inputs have the same message bandwidth W , the commutator should rotate at the rate $f_s \geq 2W$. This ensures that successive samples from any one input are spaced by:

$$T_s = \frac{1}{f_s} \leq \frac{1}{2W}$$ -- (6.1)

The time interval T_s containing one sample from each input is called *a fra*me.For M input channels, the pulse-to-pulse spacing within a frame is given by Eq. (6.2):

$$\frac{T_s}{M} = \frac{1}{Mf_s} \quad\text{--} \quad (6.2)$$

The total number of pulses per second, r, will be as per Eq. (6.3):

$$r = Mf_s \geq 2MW \quad\text{--} \quad (6.3)$$

This is referred to as the pulse rate or the signaling rate of the TDM signal. A primitive TDM system used mechanical switching to generate multiplexed PAM. However, almost all practical TDM systems employ electronic switching as in Fig. 6.7. Fig. 6.8 shows the timing diagram for TDM commutator.

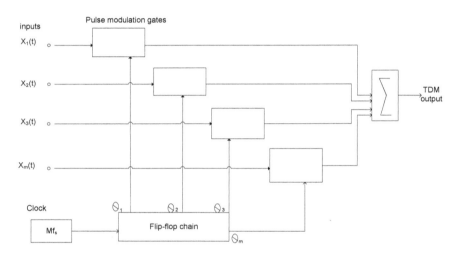

Fig. 6.7. Electronic commutator for TDM

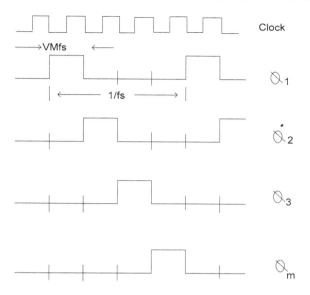

Fig. 6.8. Timing diagram for TDM commutator

Besides PAM, other types of pulse modulation may be used. Pulse modulation gates process the individual inputs to form the TDM outputs. The gate control signals come from a flip-flop chain driven by a digital clock at frequency Mf_s. The de-commutator would have a similar structure.

Synchronization is carefully needed regardless of the pulse modulation that is used between commutator and de-commutator. Synchronization is a critical consideration in TDM, because each pulse must be distributed to the correct output line at the appropriate time.

6.3.1.1 Brute-force synchronization

The technique devotes one time slot per frame to a distinctive marker pulse or non-pulse. Fig. 6.9 shows the Brute-force synchronization.

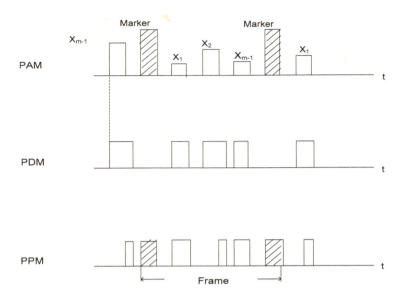

Fig. 6.9. Brute-force synchronization

The markers establish the frame frequency f_s at the receiver. The number of signal channels is reduced by 1 to $M-1$. The complete transmitter diagram with reduced signal channels is shown in Fig. 6.10:

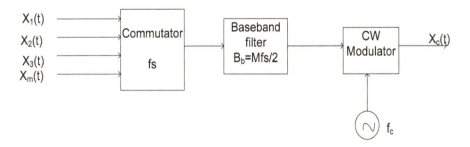

Fig. 6.10 Transmitter with reduced signal channels

A low-pass baseband filter has bandwidth given by Eq. (6.4)

$$B_b = \frac{1}{2}r = \frac{1}{2}Mf_s \quad \text{...} \quad (6.4)$$

The interlaced sample spacing equals $\dfrac{1}{Mf_s}$

If baseband filtering is employed and if the sampling frequency is close to the Nyquist rate, $f_{s\,min} = 2W$ for the individual inputs, then the transmission bandwidth for PAM/SSB becomes as in Eq. (6.5)

$$B_r = \frac{1}{2}.Mx2W = MW \quad\text{--}\quad (6.5)$$

It has been assumed that all inputs have the same bandwidth, but this assumption is not essential and would be unrealistic for the important case of analog data.

A typical telemetry system has a main multiplexer plus sub-multiplexers arranged to handle 100 or more data channels with various sampling rates.

Let us consider TDM telemetry with 5 data channels needed with minimum sampling rates as in Table 6.2:

Table 6.2. TDM telemetry with 5 data channels

Data channel	Minimum sampling rate
$x_1(t)$	3.0kHz
$x_2(t)$	0.7kHz
$x_3(t)$	0.5kHz
$x_4(t)$	0.3kHz
$x_5(t)$	0.2kHz

If a 5 channel multiplexer with $f_s = 3kHz$ for all channels brings up the TDM signaling rate $r = 5x3kHz = 15kHz$. This implies that the rate of $15kHz$, excluding synchronization markers, will be needed. Fig. 6.11 represents a TDM multiplexing operation with efficient scheme

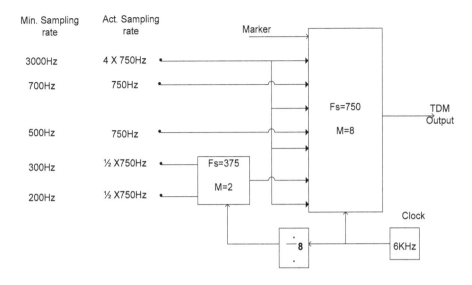

Fig. 6.11. TDM multiplexing operation with efficient scheme

An 8-channel main multiplexer with $f_s = 0.75kHz$ and 2-channel sub-multiplexer with $f_s = 0.375kHz$ is considered. The samples of $x_4(t)$ and $x_5(t)$ will appear in alternate frames. $x_1(t)$ samples appear in 4 equi-spaced slots within each frame. Therefore, the total output signaling rate $= 8x0.75kHz = 6kHz$.

6.4 Comparison of TDM and FDM

While the individual TDM channels are assigned to distinct *time slots*, they are jumbled together in the frequency domain. The FDM channels are assigned to distinct *frequency slots*, but jumbled together in the time domain.

It is advantageous to use TDM over FDM because: (i) TDM involves simpler instrumentation; (ii) TDM is invulnerable to the usual causes of crosstalk in FDM; (iii) The use of sub-multiplexers allow a TDM system to accommodate different signals whose bandwidths or pulse rates may differ by more than an order of magnitude; (iv) TDM may or may not be advantageous when the transmission medium is subject to *fading*; and (v) Rapid wideband fading might strike only occasional pulse in a given TDM channel, whereas all FDM channels would be affected. Nevertheless, slow narrowband fading wipes out all the TDM channels, whereas it might hurt only one FDM channel.

Chapter Seven

Analog Pulse Modulation

7.1 Introduction

Periodic sample values can be used to adequately describe a message waveform and can be transmitted using analog pulse modulation. Pulse parameters suitable for modulation include; (i) amplitude—for pulse amplitude modulation (PAM); (ii) duration—for pulse duration or width or length modulation (PDM or PWM or PLM); and (iii) position—for pulse position modulation (PPM);

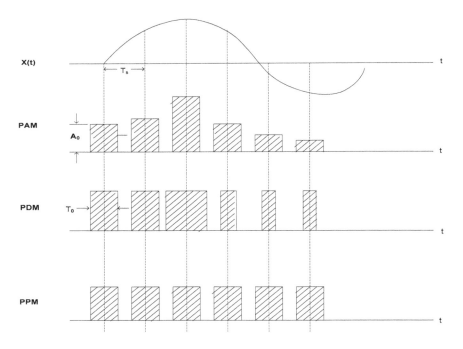

Fig. 7.1. PAM, PDM, PPM

The modulated pulse parameter—amplitude, duration, or position—varies in direct proportion to the sampled values of $x(t)$. A modulated pulse train has significant *dc* content. This means that the bandwidth required to preserve the pulse shape exceeds the message bandwidth. It is know that we hardly encounter a single-channel telecommunications system with PAM, PDM, or PPM. However, analog pulse modulation has a major role in TDM, telemetry, and so on.

7.2 What is pulse modulation?

It is a system in which continuous waveforms are sampled at regular intervals. Information regarding the signal is transmitted only at the sampling times, together with any synchronizing pulses that may be required.

At the receiver, the original waveforms may be reconstructed from the information regarding the samples, if they are taken frequently enough! The resulting receiver output can have negligible distortion.

Analog and digital are the two broad categories of PM—analog PM—infinitely variable sample amplitudes and digital PM—the nearest pre-determined level of sample amplitude. Fig. 7.2. shows the classes of analog and digital modulations whereby under digital there pulse code modulation (OCM and delta modulation. Pulse amplitude modulation (PAM) and pulse time modulation belongs to analog modulation.

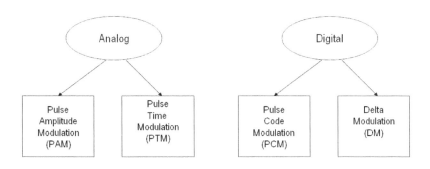

Fig. 7.2. Classes of analog and digital modulations

All the modulation systems have sampling in common, but they differ in the manner of indicating the sample amplitude. Pulse-amplitude modulation and pulse-time modulation corresponds roughly to AM and FM.

7.2.1 Pulse-amplitude modulation (PAM)

The simplest form of PM is given in the Fig. 7.3.

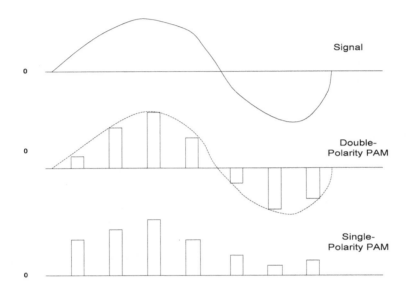

Fig. 7.3. Illustration of PM—double-polarity PAM and single-polarity PAM

PAM is a pulse modulation system in which the signal is sampled at regular intervals, and each sample is made proportional to the amplitude of the signal at the instant of sampling. The pulses are then sent by either wire or cable or else are used to modulate carrier. In the case of single-polarity PAM, a fixed dc level is added to the signal, to ensure that the pulses are always positive. There is, however, an advantage of having the ability to use constant-amplitude pulses. This gives a reason why PAM is infrequently used.

It is very easy to generate and demodulate PAM. The signal to be converted to PAM is fed to one input of an *AND gate*. Pulses at the sampling frequency are applied to the other input of the *ANG gate* to open it during the wanted time interval. The output pulses are then passed through a pulse-shaping network, which gives them flat-tops.

7.2.2 Pulse-time modulation (PTIM)

The pulse has constant amplitude. One of their timing characteristics is varied, being made proportional to the sampled signals amplitude at that time instant. The varied characteristics may be the width of the pulse or the position of the pulse or the frequency of the pulse.

Three types of PTM are possible—pulse width modulation (PWM) or (PDM); pulse position modulation (PPM); and pulse frequency modulation (PFM)

7.2.3 Pulse-width/division modulation (PWM/PDM)

Pulse-width modulation (PWM) or Pulse-duration modulation (PDM) or pulse-length modulation (PLM) adopts the technique that keeps the pulse at fixed amplitude and starting time of each pulse, but the width is proportional to the amplitude. A negative pulse width is not possible as it would make the pulse end before it began.

The disadvantage is that varying width leads to varying power content while the advantage is that: PWM works even if synchronization between transmitter and receiver fails.

7.2.4 Pulse-position modulation (PPM)

The amplitude and width of the pulses are kept constant. The position of each pulse is varied by each instantaneous sampled value. Fig. 7.4 shows an example of PWM waveforms

The advantage is that it requires constant transmitter power output while the advantage is that depends on transmitter-receiver synchronization.

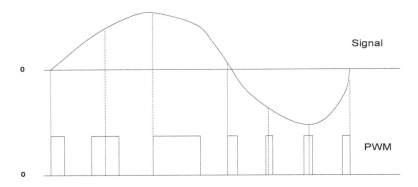

Fig. 7.4. Example of PWM waveforms

Using the waveforms, as an assignment, a reader can describe how to generate and demodulate PWM and PPM?

7.3 Sampling Theorem

Sampling theorem statement: "If the sampling rate in any pulse modulation system exceeds twice the maximum signal frequency, the original signal can be reconstructed in the receiver with minimal distortion.". The sampling theorem is used in practice to determine minimums sampling speeds.

Let us look into an example of pulse modulation for speech audio frequency range which is 300~3400Hz. The sampling rate will be 8000 samples per second. This is almost a world-wide standard. The pulse rate is greater than twice the highest audio frequency. Therefore, the sampling theorem is satisfied and the system is optimistically assumed to be free from sampling error.

Chapter Eight

Pulse-Code Modulation

8.1 Introduction

Starting with the technique, the combined operations of *sampling* and *quantizing* generate a quantized PAM waveform which is a train of pulses whose amplitude are restricted to a number of discrete magnitudes. Each quantized level can be represented by a code number. The code number can be converted into its binary arithmetic which can be the digits of the binary representation of the code number that are transmitted as pulses, normally in binary. This is what we refer to as pulse-code modulation (PCM).

Each one of the digits, almost always in binary code, represents the approximate amplitude of the signal sample at that instant. The approximation can be made as close as desired, but it is always just an approximation.

8.2 Principle of PCM

Doing PCM codes starts by making sure that the total amplitude range which the signal may occupy is divided into a number of standard levels. Fig. 8.1 illustrates the division.

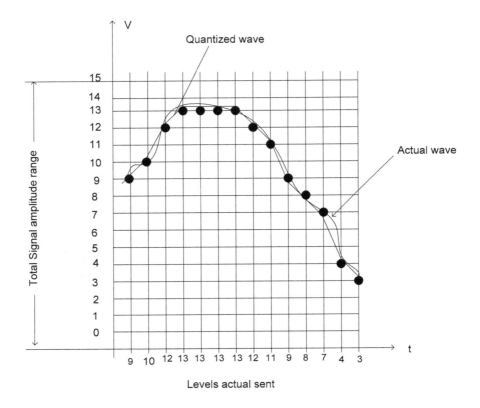

Fig. 8.1. Principle of PCM

Since the levels are transmitted in a binary code, the actual number of levels is put to a number equal to the power of 2. In practice, the telecommunications systems use as many as 128 levels. For the case of 16 levels, that is 2^4, we must be ensured of 4 binary places for transmission.

Let us consider an example of an actual signal with amplitude of 6.8 V, The signal is not sent as a 6.8V pulse as it might have been in PAM, nor as a 6.8µs-wide pulse as in PWM. It will rather simply be sent as the digit 7 which in binary digits means 0111 or 0PPP where P corresponds to pulse and 0 to no pulse. Actually, the code is sent as a binary number back-to-front, that is, as 1110 or PPP0. This makes the demodulation mechanism easier.

Therefore, PCM is about sampling, then quantizing, then coding, and then sending of the coded strings. Provided that sufficient quantizing levels are used, the results can not be distinguished from that of analog transmission. A supervisory bit or signalling bit is generally added to each code group representing a quantized sample. Each group of pulses denoting a sample is called *a word* and it is expressed by means of (n+1) bits where 2^n is the chosen number of standard levels.

PCM can be viewed as a digital transmission system with an analog-to-digital converter (ADC) at the input and a digital-to-analog converter (DAC) at the output. PCM performance as an analog telecommunications system depends on the quantization noise introduced by ADC and the digital error probability.

8.3 PCM generation and reconstruction

A PCM generating system can be represented by the functional blocks as shown in Fig. 8.2.

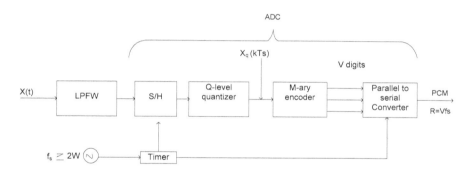

Fig. 8.2. PCM generation and reconstruction

In the systems, it is shown that the analog input $x(t)$ is low-pass filtered and sampled to $x(kT_s)$. The quantizer, then, rounds off the sample values to the nearest discrete value in a set of q quantum levels, that is, $x_q(kT_s)$. It must be clear that $x_q(kT_s)$ are discrete in time which clearly signifies the sampling process, while $x_q(kT_s)$ are discrete in amplitude to reflect the quantizing stage.

8.3.1 Relationship between $x(kT_s)$ and $x_q(kT_s)$

Let a voltage waveform $x(t)$ normalized such that it can be expressed as in Eq. (8.1):

$$|x(t)| \leq 1 Volt \text{--- (8.1)}$$

Uniform quantization divides the *2 Volts* peak-to-peak range into q equal steps of height $\dfrac{2}{q} Volts$ .each as described in Fig.

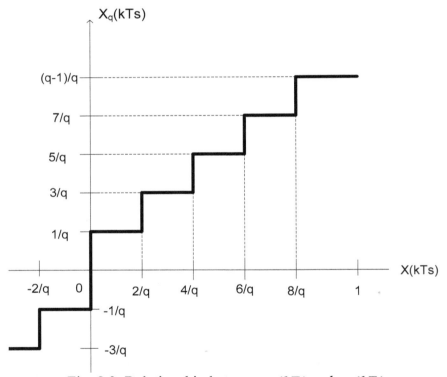

Fig. 8.3. Relationship between $x(kT_s)$ and $x_q(kT_s)$

In terms of the quantization characteristic, we have quantum levels at $\pm\dfrac{1}{q}, \pm\dfrac{2}{q}, \pm\dfrac{3}{q}, ..., \pm\dfrac{(q-1)}{q}$

A quantized value such as $x_q(kT_s) = \dfrac{5}{q}$ corresponds to any value in $\dfrac{4}{q} < x(kT_s) < \dfrac{6}{q}$ range. Encoder translates the quantized samples into digital code words. There are M^v possible M-ary code-words with v digits per word. Unique encoding requires that for q different quantum levels, Eq. (8.2) must hold:

$$M^v \geq q \quad\text{.. (8.2)}$$

In addition, M, V, and q should satisfy the equality in Eqs. (8.3) and (8.4):

$$q = M^v \quad\text{.. (8.3)}$$

This means that

$$v = \log_M q \quad\text{... (8.4)}$$

For binary PCM, it implies that:

$$q = 2^v \quad\text{... (8.5)}$$

The PCM generator acts as an ADC performing at:

$$f_s = \frac{1}{T_s} \quad\text{... (8.6)}$$

Timer coordinates the sampling and parallel-to-serial readout. Each encoded sample is represented by a $v - digit$ output word. This means the signaling rate, r, is given by Eq. (8.7):

$$r = vf_s \quad\text{... (8.7)}$$

where $f_s \geq 2W$

The bandwidth needed for PCM baseband transmission is expressed in Eq. (8.8):

$$B_T \geq \frac{1}{2}r = \frac{1}{2}vf_s \geq \frac{1}{2}v.2W = vW \quad \text{...........................} \quad (8.8)$$

A fine-grain quantization for accurate reconstruction of the message waveform requires that $q >> 1$. This increases the B_T by the factor $v = \log_M q$ times the message bandwidth W.

8.3 A PCM receiver

Fig. 8.4 shows a schematic diagram of a PCM receiver with a DAC block which converts the regenerated PCM codes into $x_q(t)$ signal samples just after having passed through a sample and hold (S/H) circuit.

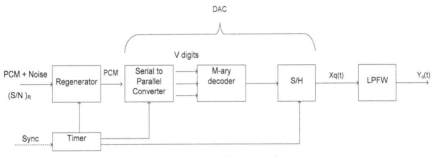

Fig. 8.4. A PCM receiver

The received signal may be contaminated by noise. If $(S/N)_R$ is sufficiently large, the regeneration yields a clean and nearly errorless waveform. The output signal $y_o(t)$ differs from the message $x(t)$ to the extent that the quantized samples differ from the exact sample values $x(kT_s)$. Perfect message reconstruction is almost impossible in PCM, even when random noise has no effect. The ADC at the transmitter introduces permanent errors that appear at the receiver as quantization noise in the reconstructed signal.

As an assignment, we must try to understand the PCM hardware implementation and be able to calculate appropriate values of v, q, and f_s for a binary channel with $r_b = 36,000 bits /\sec.$ when it is available for PCM voice transmission assuming that $W=3,200kHz$.

8.4 Quantization noise

Let us consider the impulse reconstruction model given in Fug.:

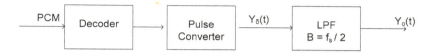

Fig. 8.5. Impulse reconstruction model

A pulse converter generates the weighted impulse train given by Eq. (8.9):

$$y_\delta(t) = \sum_k \left[x(kT_s) + \varepsilon_k \right] \delta(t - kT_s) \quad\text{.............................} \quad (8.9)$$

where ε_k is the quantization error, given by Eq. (8.10):

$$\varepsilon_k = x_q(kT_s) - x(kT_s) \quad\text{.................................} \quad (8.10)$$

Low-pass filtering with $B = \dfrac{f_s}{2}$ yields the final output given by Eq. (8.11):

$$y_o(t) = x(t) + \sum \varepsilon_k Sinc(f_s - k) \quad\text{..........................} \quad (8.11)$$

When q is large enough, ε_k will be uncorrelated and independent of $x(t)$.

Let identify $\overline{\varepsilon_k^2}$ as the mean-square quantization noise such that, by rounding-off quantization, Eq. (8.12) is valid:

$$\left| \varepsilon_k \right| \leq \frac{1}{q} \quad\text{.......................................} \quad (8.12)$$

If the quantization error has zero mean value and a uniform *pdf* over $-\dfrac{1}{q} \le \varepsilon_k \le \dfrac{1}{q}$, then the quantization noise power is given by Eq. (8.13):

$$\sigma_q^{\ 2} = \overline{\varepsilon_k^{\ 2}} = \frac{1}{\left(\frac{2}{q}\right)} \int_{-\frac{1}{q}}^{\frac{1}{q}} \varepsilon^2 d\varepsilon = \frac{1}{3q^2} \quad\text{(8.13)}$$

This implies that the quantization noise decreases as the number of levels increases.

8.5 PCM performance

The measures to consider include the destination signal power given by Eq. (8.14) and the quantization noise power, $\sigma_q^{\ 2}$

$$S_D = \overline{x^2} = S_x \le 1 \quad\text{(8.14)}$$

The destination signal-to-noise ratio is given as in Eq. (8.15):

$$\left(\frac{S}{N}\right)_D = \frac{S_x}{\sigma_q^{\ 2}} = \frac{S_x}{\left(\frac{1}{3q^2}\right)} = 3q^2 S_x \quad\text{(8.15)}$$

Setting $q = 2^v$, we have Eq. (8.16) or Eq. (8.17):

$$\left(\frac{S}{N}\right)_D = 3.(2^v)^2 S_x = 3.2^{2v} S_x \quad\text{(8.16)}$$

In decibels we have,

$$\left(\frac{S}{N}\right)_{D,dB} = 10\log(3.2^{2v} S_x) \quad\text{(8.17)}$$

$$\le 4.8 + 6.0v \ dB$$

The upper bound holds when $S_x = 1$.

A typical example is a voice telephone PCM system with $v=8$ that has $\left(\dfrac{S}{N}\right)_{D,dB}$ computed as in Eq. (8.18):

$$\left(\frac{S}{N}\right)_{D,dB} \leq 4.8 + 6.0x8 = 52.8dB \quad \text{------------------------------} \quad (8.18)$$

As an assignment, let us explain *crest factor*, $\left|x(t)\right|_{max}/\sigma_x$. and give an explanatory example. Descriptions of *companding* and *companding curves* for PCM can also be found from other literature.

8.6 Advantages of PCM and its applications

It is known that PCM is so awesome, but still other modulation systems are in uses. There are basically three main reasons for this. First, the other systems came first while PCM came in 1937. Secondly, PCM requires very complex encoding and quantizing circuitry. Thirdly, PCM requires a larger bandwidth compared to analog systems.

For our own interest, PCM was invented by Alex H. Reeves in Great Britain in 1937, but complexity prevented its immediate use. It started to be in use by the end of WWII and it was actually used for telephony after 1960. PCM requires complex circuitry, but multiplexing equipment is very cheap. Repeaters do not have to be placed so close together as PCM tolerates much worse SNR. Finally, in PCM, bandwidth requirement is no longer as serious as it had earlier been because we now have large bandwidth fibre optic systems in use.

The Bell T1 digital transmission system in use in North America, for example, uses 24 PCM channels, TDM, 8 bits/sec/channel which means 24x8+1=193bits with 1 being sync signal. The sampling rate is 8000/sec and, hence, it gives 8000x193=1,544,000bits per second that will be sent.

To conclude, it has been know that PCM has uses in space communications. For example, it was amazing that, with

MARINER IV in 1965, PCM transmitted the first pictures of Mars, almost 200,000,000km away with transmitting power of only 10W. It is the same PCM we are talking about that was used. No other system would have done the job as great as PCM did.

Chapter Nine

Data Transmission

9.1 Introduction

By general definition, data transmission is the *sending* of data from one place to another by means of *signals* over a *channel*. This is what exactly any telecommunications system would be pleased to do.

9.2 Base-band data transmission

Most physical layer transmission systems rely on baseband transmission. Let us consider a simple analog baseband transmission system as shown in Fig. 9.1.

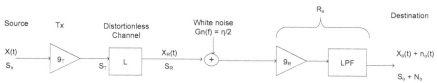

Fig. 9.1. Analog baseband transmission system

The information source generates a message waveform $x(t)$ which is to be produced at the destination. Let the message bandwidth is W and that $x(t) \cong 0$ for $|f| > W$. We assume that the channel is distortion-less over the message band and, so, Eq. (9.1) holds.

$$x_D(t) = Kx(t - t_d) \quad\text{-- (9.1)}$$

where K signifies the total amplification and t_d is the time delay of the system.

Let us represent the average signal power generated at the source by Eq. (9.2):

$$S_x \overline{\underline{\Delta x^2}} \quad\text{---------------------------------------} \quad (9.2)$$

The transmitter and the receiver act as amplifier with power gains g_T and g_R. The two are to compensate for the transmission loss L. Therefore, the transmitted signal power, the received signal power, and the destination signal power are related as in Eqs. (9.3) and (9.4):

$$S_T = g_T \overline{x^2} = g_T S_x \quad\text{------------------------} \quad (9.3)$$

$$S_R = \overline{x_R^2} = \frac{S_T}{L} \quad\text{--------------------------} \quad (9.4)$$

The filter, here, has the crucial task of passing the message while reducing the noise at the destination. The filter should reject all noise frequency components that fall outside the message band and hence we find the need for an ideal LPF with bandwidth $B = W$.

If the receiver has power gain g_R and noise equivalent bandwidth B_N, the destination noise power can be expressed as in Eq. (9.5):

$$N_D = g_R \eta B_N \quad\text{----------------------------} \quad (9.5)$$

The resulting destination noise power will be given in Eq. (9.6):

$$N_D = g_R \eta W \quad\text{-------------------------------} \quad (9.6)$$

The receiver gain g_R amplifies signal and noise equally. As g_R cancels out, this results into Eq. (9.7):

$$(S/N)_D = \frac{g_R S_R}{g_R \eta W} = \frac{S_R}{\eta W} \quad\text{-----------------} \quad (9.7)$$

$(S/N)_D$ is given in terms of three fundamental system parameters: S_R, η (receiver side), and W, while ηW is the noise power in the message band. This means that a wideband signal suffers more from noise contamination than a narrowband signal.

In decibels, we will have Eq. (9.8):

$$(S/N)_{D,dB} = 10\log_{10}\left(\frac{S_R}{\eta W}\right) = 10\log_{10}\left(\frac{S_R}{kT_N W}\right) \quad\text{(9.8)}$$

Expressing the signal power in mW or dBm, then we will have Eq. (9.9):

$$(S/N)_{D,dB} = S_R dBm + 174 - 10\log_{10}\left(\frac{T_N}{T_o}W\right) \quad\text{(9.9)}$$

where

$$\eta = kT_N = kT_o\left(\frac{T_N}{T_o}\right) = 4x10^{-21}\left(\frac{T_N}{T_o}\right) \; W/Hz \quad\text{(9.10)}$$

Table 9.1 shows typical transmission requirements for selected analog signals in practical use.

Table 9.1. Typical transmission requirements for selected analog signals

SNo.	Signal type	Frequency range	SNR in dB
1	Barely intelligible voice	0.5 ~ 2kHz	5 ~ 10
2	Telephone-quality voice	0.2 ~ 3.2kHz	25 ~ 35
3	AM broadcast-quality voice	0.1 ~ 5kHz	40 ~ 50
4	High-fidelity audio	0.02 ~ 20kHz	55 ~ 65
5	Television video	0.06 ~ 4.2MHz	45 ~ 55

The upper limit of the frequency range is the nominal message bandwidth, W.

The value $(S/N)_D$ does not depend on the receiver gain, which only serves to produce the desired signal level at the output. $(S/N)_D$ will be affected by any gains or losses that enter the picture before the noise has added. For example, if we substitute $S_R = \dfrac{S_T}{L}$, then we get Eq. (9.11)

$$(S/N)_D = \left(\frac{S_T}{L}\right)/\eta W = \frac{S_T}{L\eta W} \quad \text{(9.11)}$$

$(S/N)_D$ is directly proportional to the transmitted power S_T and inversely proportional to the transmission loss L. This has a significant conclusion in the design of telecommunications systems.

When all the parameters in Eq. (9.11) are fixed and $(S/N)_D$ turns out to be too small, we see the importance of using repeaters, as described in Fig. 9.2, to improve the system's performance.

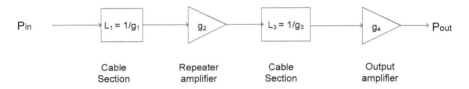

Fig. 9.2. Transmission system with repeater amplifier and cable sections

Considering the transmission system in Fig. 9.2, let that the transmission path is divided into m equal sections, each having loss L_1, we will have $(S/N)_1$ and $(S/N)_D$ expressed as in Eq. (9.12) and Eq. (9.13):

$$(S/N)_1 = \frac{S_T}{L_1\eta W} \quad \text{(9.12)}$$

and

$$(S/N)_D \cong \frac{1}{m}(S/N)_1 = \frac{L}{mL_1}\left(\frac{S_T}{L\eta W}\right) \text{\dotfill} (9.13)$$

Eq. (9.13) has shown a potential improvement by a factor of $\frac{L}{mL_1}$.

All of the above results have been based on distortion-less transmission, additive white nose, and ideal filtering.

It must be noted that if the noise bandwidth is greater than the message bandwidth, W, that is $B_N > W$, then $(S/N)_D$ will be reduced by the factor W/B_N.

System non-linearities also cause the reduction of $(S/N)_D$. However, non-linear companding may end up in to a net improvement.

Taking an example, let L for a cable system is equal to $140dB = 10^{14}$; $T_N = 5T_o$; $W = 20kHz$, representing high-fidelity audio transmission; and let $(S/N)_D \geq 60dB$. Then, we start from Eqs. (9.14) and (9.15):

$$(S/N)_{D,dB} = 10\log_{10}(S_R/kT_N W) \text{\dotfill} (9.14)$$

$$= S_R dBm + 174 - 10\log_{10}\left(\frac{T_N}{T_o}W\right) \text{\dotfill} (9.15)$$

From the specifications, we can deduce that: $S_R dBm + 174 - 10\log_{10}(5x20x10^3) \geq 60dB$.

Therefore,

$$S_R \geq -64dBm \cong 4x10^{-7} mW \text{\dotfill} (9.16)$$

It is surprising that the transmitted power $S_T = LS_R \geq 4x10^7 mW = 40,000W$ which we can not even try to put on a signal transmission cable. Instead, we opt to insert a repeater at the midpoint so that $L_1 = 70dB = 10^7$.

Therefore, the resulting improvement factor is:

$$\frac{L}{mL_1} = \frac{10^{14}}{2x10^7} = 5x10^6$$

This reduces the transmitted power equivalent to $S_T \geq \frac{4x10^7}{5x10^6} = 8mW$ which is a much more realistic value in real practice.

To provide a margin of safety, S_T may be taken in the range of $10 \sim 20mW$ and that is what we practical do.

9.3 Error control coding

Errors enter the data stream during transmission and are caused by noise and transmission system impairments. Errors compromise the data and in some case render it useless. Therefore, there is a need to detect and correct transmission errors. Error correction process normally results in to an increase in the number of bits per second to be transmitted which increases the cost of transmission.

The tolerance the data user has for errors will decide which error control system is appropriate for the transmission circuit being used for the user's data.

9.3.1 Error detection

9.3.1.1 Parity-check codes

It is a popular form of error detection employing redundancy. A parity-check bit is added to each character or code group. The logical process produces the parity-check codes.

The most common techniques involves adding the *1s* in each character block code and append a *1* 0r a *0* as required to obtain an odd or even total. For odd parity system, we add a *1* if addition of the *1s* in the block sum is odd.

9.3.1.2 Constant-ratio codes

When the group is received, the receiver will be able to determine that an error has occurred if the ratio of *1s* to *0s* has been altered. If an error is detected, a NAK—not acknowledged—response is sent and the data word is repeated and retransmitted.

The limitation of this scheme is that an odd number of errors will be detected, but an even number may go undetected.

If *T=total bits* and *M = number of 1s*, then the number of combinations will be given by Eq. (9.17):

$$The\ number\ of\ combinations = \frac{T!}{M!(T-M)!} \quad\quad (9.17)$$

Let us take the common examples elaborated in other literature:

(a) For the case of the 2-out-of-5 code, the number of combinations are given as in Eq. (9.18):

$$The\ number\ of\ combinations = \frac{5!}{2!(5-2)!} = \frac{120}{2x6} = 10 \quad\quad (9.18)$$

(b) For the case of the 4-out-of-8 code, the number of combinations are given as in Eq. (9.19):

The number of combinations

$$= \frac{8!}{4!(8-4)!} = \frac{8x7x6x5}{4!} = \frac{8x7x6x5}{4!} = 70 \quad\quad (9.19)$$

This shows improved error detection but reduced efficiency. In principle, the code depends on the rates of *1s* to *0s* in each code group to indicate that errors have occurred.

9.3.1.3 Redundant codes

In this technique, information additional to the basic data is sent. For example, the technique is used to transmit the information twice and compare the two sets of data to see that they are the same. Statistically, it is very unlikely that a random error will occur a second time at the same place in the data. If a discrepancy is noted between the two sets of data, an error is assumed and the data is caused to be retransmitted. Otherwise, error-free transmission is assumed to have taken place.

The problem of this scheme is that it introduces 100% redundant and, therefore, it is very inefficient. In the above case, the efficiency is obtained as in Eq. (9.20):

$$Efficiency = \frac{Information\ bits}{Total\ bits} = 50\% \dots\dots\dots\dots\dots\dots\dots\ (9.20)$$

Eq. (9.20) suggests that in a system with no redundancy, there is 100% efficiency. Therefore, most systems of error detection will fall between *50%* and *100%* of efficiency.

9.3.2 Error correction

Practically, detecting error an important step, but it is of little use unless methods are available and implementable for the correction of the detected errors. An important aspect of data transmission remains to be error correction mechanism. Error correction coding is more sophisticated than error detection coding. However, they are both important.

9.3.2.1 Retransmission

Data or packet retransmission is the most popular method of error correction where the retransmission of the erroneous information is done through an automatic system. It is needed to have automatic request for repeat (ARQ). This is the system in common and practical use.

In addition, we have what is called positive acknowledgement or negative acknowledgement method, in short ACK or NAK method.

9.3.2.2 Forward error correction codes (FEC)

For efficiency, error correction at the receiver without retransmission of erroneous data is preferred. Codes which permit correction of errors by the receive station without retransmission are called forward-error-correcting codes. The requirements for this technique include the provision for sufficient redundancy that need to be included.

To understand this technique, let us consider a typical FEC system as diagrammed in Fig. 9.3.

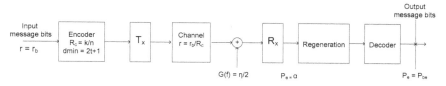

Fig. 9.3. A typical FEC system

Message bits come from an information source at rate r_b. The encoder takes blocks of k message bits and constructs (n,k) block code with code rate $R_c = k/n < 1$. The bit rate, r, on the channel must be greater than r_b, That is r and r_b, must be related as in Eq. (9.21):

$$r = \left(\frac{n}{k}\right) r_b = r_b R_c \text{---(9.21)}$$

The minimum distance is given by Eq. (9.22):

$$d_{\min} = 2t + 1 \leq n - k + 1 \text{----------------------------------(9.22)}$$

If E_b is the average energy per message bit, then the average energy per code bit is $R_c E_b$ and the ratio of bit energy to noise density is given by Eq. (9.23):

$$\gamma_c \triangleq \frac{R_c E_b}{\eta} = R_c \gamma_b \text{------------------------------------(9.23)}$$

The performance criterion is the probability of output message bit error, P_{be}, to distinguish it from the word error probability, P_{we}. The code connects up to t errors per word. Therefore, the probability of a decoding word error is upper bounded by an expression given by Eq. (9.24).

$$P_{we} \leq \sum_{i=t+1}^{n} P(i,n) \cong P(t+1,n) \cong \binom{n}{t+1}\alpha^{t+1} \quad \text{(9.24)}$$

An uncorrected word has typically $t+1$ bit errors. On the average, there will be $(k/n)(t+1)$ message bit errors per uncorrected word. The remaining errors being in check bits.

When Nk bits are transmitted in $N \gg 1$ words, the expected total number of erroneous message bits at the output is given in Eq. (9.25):

$$(k/n)(t+1)NP_{we} \quad \text{(9.25)}$$

This implies that P_{be} will be expressed as in Eq. (9.26):

$$P_{be} = \frac{t+1}{n}P_{we} \cong \binom{n-1}{t}\alpha^{t+1} \quad \text{(9.26)}$$

The transmission error probability is given by Eq. (9.27):

$$\alpha = Q\left(\sqrt{2\gamma_c}\right) = Q\left(\sqrt{2R_c\gamma_b}\right) \quad \text{(9.27)}$$

$$\cong \left(4\pi R_c\gamma_b\right)^{-1/2} e^{-R_c\gamma_b} \quad \text{(9.28)}$$

with $R_c\gamma_b \geq 5$.

Assuming $\alpha \ll 1$, then the final result for the output error probability of the FEC system becomes as the one expressed in Eq. (9.29) or Eq. (9.30):

$$P_{be} = \binom{n-1}{t}\left[Q\left(\sqrt{2R_c\gamma_b}\right)\right]^{t+1} \quad \text{(9.29)}$$

$$\cong \binom{n-1}{t}\left(4\pi R_c\gamma_b\right)^{-(t+1)/2} e^{-(t+1)R_c\gamma_b} \quad \text{(9.30)}$$

For un-coded transmission we have P_{ube} given as in Eq. (9.31):

$$P_{ube} = Q\left(\sqrt{2\gamma_b}\right) \cong \left(4\pi\gamma_b\right)^{-\frac{1}{2}} e^{-\gamma_b} \quad\text{---} \quad (9.31)$$

Part II
Digital Telecommunications

Chapter Ten

Introduction to Digital Telecommunications

10.1 Introduction digital technique

Digital technology is a branch of electronics and communications which utilizes discontinuous signals, that is, signals which appear in discrete steps rather than having the continuous variations characteristics of analog signals.

Digital techniques must have the ability to construct unique codes to represent different items of information. Digital technology is of particular importance when information is to be: gathered, stored, retrieved, and evaluated. Digital processing is used so widely because it provides economical and rapid manipulation of data.

It has been taken as a surprise that everything is in struggle to go digital. Military telecommunications systems, commercial telecommunications systems, educational telecommunications systems, research telecommunications systems, health telecommunications systems, and so on, are all in the process of going digital.

The inspiration for that has many reasons. These include: (i) the ease with which digital signals are regenerated—the pulses can be reborn or regenerated using regenerative repeaters; (ii) digital circuits are less subject to *distortion* and *interference* than are analog circuits—they normally operate on one of two states—*on or off* to be meaningful; (iii) high signal fidelity techniques are possible through error detection and correction, but not possible with analog; (iv) digital circuits are more reliable and can be produced at a lower cost than analog circuits; (v) digital hardware leads to more flexible implementation; (vi) the combining of digital signals using TDM is simpler than the combining of analog signals using FDM—different types of digital signals—data, telegraph, telephone, television, audio, images, and videos—can be treated as identical signals in transmission and switching—a bit is a bit—a packet can include each type of information at the same time; (vii) protection against interference and jamming or good encryption and privacy are better done with digital techniques; and (viii) much data communication is from computer to computer or from digital instruments or terminal to computer—best served by digital telecommunications links.

The cost of digital telecommunications systems is more based on the nature of the processes that take place within the systems. As well known, digital systems are very signal-processing intensive and, hence, a significant share of the resources needs to be allocated to the task of *synchronization* at various levels.

The main and the pronounceable disadvantage of digital telecommunications systems is a non-graceful degradation of performance. When SNR drops below a certain threshold, the quality of service can change from very good to very poor. In some cases, this is what brings a false alarm due to wrong detection, in radar systems, for example.

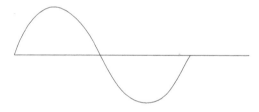

Fig. 10.1. Sine wave—an example of an analog signal

Fundamentally, digital data must be compared with analog data in order to understand the distinction between the two. The simplest unambiguous example of analog signal is a sine wave which is continuous with respect to time. Fig. 10.1 shows this Sine wave with amplitude being left open to any metric of signal strength. The horizontal axis is the time measure. Something to be noted is that, a signal does not need to be smooth like Sine or Cosine wave to qualify as analog. The continuity with respect to time is the primary and basic qualification.

Digital signal, on the other side, is a representation of data as a series of digits, example a number. A signal is represented by codes which approximates the actual values. The binary digit system is based on the binary numbering system where only *0* and *1* make the alphabet of the codewords.

To understand the theme clearly, let us take an example of a binary representation of the Sine wave signal.

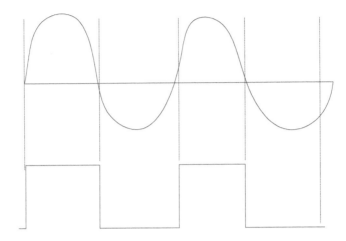

Fig. 10.2. Binary representation of a Sine wave

Fig. 10.2 shows a schematic binary representation of a Sine wave with an option of *1* for the upper parts of the cycles and *0* for the lower parts. A series of the sine-wave cycles would generate a series of pulses with spaces between them.

In summary, digital means that, the information is coded in the form of binary states of *1s* and *0s* in general practical systems in use.

10.2 Elements of a digital telecommunications system

Let us consider Fig. 10.3 which provides simplified and generalized elements of a digital telecommunications system.

The upper blocks include the formatting, the source, the source encoding, the encryption, the channel encoding, the multiplexing, the pulse modulating, the bandpass modulating, the frequency spread, and the multiple accessing sections which all together denote signal transformation from the source to the transmitter (XMT).

The lower blocks denote signal transformations from the receiver to the sink. This implies the reversing of the signal transformation or processing steps performed by the above blocks. However, it must be clear that the modulating section and the demodulating or detecting blocks represent a modem. Modem often does several of the signal processing steps. This is to say that a modem in a telecommunications system represents the brains of the system while the receiver and the transmitter represent the muscles of the system.

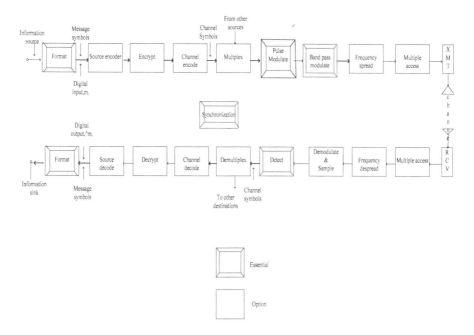

Fig. 10.3. Elements of a digital telecommunications system

For wireless applications, the transmitter consists of frequency up-conversion stage to a radio frequency (RF), a high power amplifier, and an antenna. This means that the receiver portion consists of an antenna and a low-noise amplifier (LNA). Frequency down-conversion is performed in the front end of the receiver or the demodulator.

During the processes, the information source is converted to binary digits or bits. Bits are then grouped to form digital messages or message symbols. Channel coding includes error correction coding where message symbols are transformed into code symbols.

Only formatting, modulating, demodulation or detection, and synchronization are essential for a digital telecommunications system. Modulation is the process by which message symbols or channel symbols are converted to waveforms that are compatible with the requirements imposed by the transmission channel. Subsequently, the pulse-modulation block usually includes filtering for minimizing the transmission bandwidth.

When pulse-modulation is applied to binary symbols, the resulting binary waveform is called a pulse-code modulation (PCM) waveform. When pulse-modulation is applied to non-binary symbols, the resulting waveform is called an M-ary pulse-modulation waveform.

Multiplexing and multiple access procedures combine signals that might have different characteristics or might originate from different sources, so that they can share a portion of the telecommunications resources like spectrum and time.

Frequency spreading can produce a signal that is relatively invulnerable to interference, both national and intentional—and can be used to enhance the privacy of the communicators.

Worth to understand is that, the blocks can sometimes be implemented in a different order. Nevertheless, all in all, synchronization plays a role in regulating the operation of almost every block.

We have mentioned about the role of signal processing in digital techniques. The basic signal processing functions goal at transformations and they are classified into nine groups. These are: (a) formatting and source coding—PCM, compression, and quantization; (b) baseband signaling—NRZ, RZ, PAM, PPM, and PDM; (c) bandpass signaling—PSK, FSK, ASK, CPM; DPSK, FSK, ASK, and CPM; (d) equalization—MLSE and with filters; (e) channel coding—antipodal, orthogonal, block, convolution, and Turbo; (f) multiplexing and multiple access—FDM, FDMA, TDM, TDMA, CDM, and CDMA; (g) spreading—DS, FH, TH, and Hybrid; (h) encryption—Block, data, and stream; and (i) synchronization.

10.3 Basic digital telecommunications nomenclature

Information source, textual message, character, binary digit or bit, bit stream, symbol, and digital waveform are the terms frequently used in digital telecommunications. They make an important part of the dictionary of the digital world. However, they have sometimes not been used according to the real spirit of the digital telecommunications nomenclature.

Information source refers the device producing information to be communicated by means of the digital telecommunications system. This can be analog or discrete. Analog information source can be transformed into digital source through the use of *sampling and quantization* which involves formatting and source encoding.

Textual message is a sequence of characters or symbols from a set of finite symbols or alphabet.

Character is a member of an alphabet or set of symbols for a given system. For character encoding, ASCII, EBCDIC, Hollerith, Baudot, Murray, and Morse have been common in the field of telecommunications engineering.

Binary digit or bit is the fundamental information unit for all digital systems. It is basically either a *1* or a *0*.

Bit stream is a sequence of binary digits, that is, ones and zeroes. Baseband signal ranges from dc to a finite value. That is, $0 \leq$ *spectral content of a bit stream* $<$ *a few MHz* .

Symbol or digital message is a group of *k* bits considered as a unit. It is called a message symbol.

Digital waveform is a voltage or current waveform that represents a digital symbol.

10.4 Digital versus analog performance criteria

Analog systems draw their waveforms from a continuum and hence they form infinite sets which demand a receiver to deal with an infinite member of possible wave shapes. Therefore, the figure of merit for the performance of analog telecommunications systems is a fidelity criterion, that is, SNR, percentage distortion, or expected mean-square error between the transmitted and the received waveforms.

On the other side, a digital telecommunications system transmits signals that represent digits. These digits form a finite set of a priori to the receiver. Therefore, a figure of merit for digital telecommunications system is the probability of incorrectly detecting a digit which is termed as the probability of error (P_E).

Chapter Eleven

Pulse Code Modulation (PCM)

11.1 Introduction

Together with pulse code modulation (PCM), pulse amplitude modulation (PAM) and pulse time modulation (PTM) are all the forms of pulse modulation. To remind ourselves, AM, FM, PAM, and PTM are all forms of analog communication where a signal is sent which has a characteristic that it is infinitely variable and proportional to the modulating voltage. PAM and PTM differ from AM and FM, because in PAM and PTM the signal was sampled and sent in pulse form. PCM also uses the sampling techniques, but it differs from PAM/PTM because PCM is a digital process. Instead of sending a pulse train varying continuously at one of the parameters, the PCM generator produces a series of numbers, or digits. Each one of these digits, almost always in binary code, represents the approximate amplitude of the signal sample at that instant. The approximation can be made as close as desired, but, as described earlier, it remains to be just and only an approximation to the real signal value.

Pulse code modulation (PCM) is the class of baseband signals obtained from the quantized PAM signals by encoding each quantized sample into a digital word. The source information is sampled and quantized to one of L levels, then each quantized sample is digitally encoded into an $l - bit$ codeword, such that l and L are related as in Eq. (11.1):

$$l = \log_2 L \quad\text{---} \quad (11.1)$$

For transmission, the codeword bits will then be transformed to pulse waveforms. The essential features of binary PCM are shown Fig. 11.1 where the natural samples, quantized samples, code numbers, and PCM sequences are shown for a given typical example of a signal excursion.

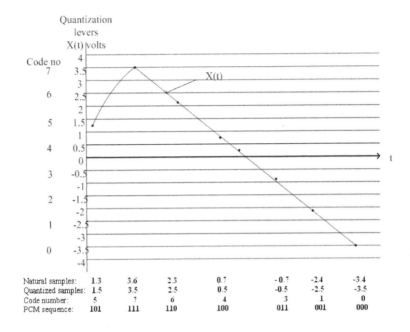

Quantization levels X(t) volts							
Natural samples:	1.3	3.6	2.3	0.7	-0.7	-2.4	-3.4
Quantized samples:	1.5	3.5	2.5	0.5	-0.5	-2.5	-3.5
Code number:	5	7	6	4	3	1	0
PCM sequence:	101	111	110	100	011	001	000

Fig. 11.1. Features of binary PCM operation in an example

For the excursions limited to the range -3.5 to $+3.5$ *Volts*, as in Fig. 11.1, with the step size set at *1 Volt,* if we do the operation in eight quantization levels at $-3.5, -2.5, -1.5, ..., +3.5$ *Volts* . The code number *0* has been assigned to the level at $-3.5 Volts$, code number 1 to the level at $-2.5 Volts$, and so on, until the level at $3.5 Volt$, that has been assigned the code number *7*. Each code number has its representation in binary arithmetic, ranging from *000* for code number *0* to *111* for code number 7. It is more convenient for the levels to be symmetrical about zero. The intervals between the levels should be equal. For a three-bit PCM sequence, each sample is assigned to one of eight levels.

Let that the analog signal is a musical passage, sampled at Nyquist rate, and let that listening this music in digital form sounds terrible, this means that the fidelity needs to be improved. What we can do is to recall the process of quantization as an approximation process contaminated by noise. Therefore, increasing the number of levels from 8 to 16 will reduce noise, but the consequences of doubling the number of levels is that each analog sample will be represented as a $4 - bit$ PCM sequence and the cost of that, in a real-time communication system, will be that the system's messages must not be delayed. This means that the transmission time for each sample must be the same, regardless of how many bits represent the sample. When there are more bits per sample, the bits must move faster which means that they must be replaced by *skinnier* bits.

Hence, the data rate will be increased and this attracts a greater transmission bandwidth and this reflects the fact that we can generally obtain better fidelity at the cost of more transmission bandwidth. Nevertheless, in some communication applications delay is permissible. The best example where delay is not a big deal if we can get the work done is the famous Galileo project where the spacecraft took almost six years between 1989 and 1995 with a mission to photograph and transmit images of the planet Jupiter. Signal delay of several minutes, or hours, or even days, would not be a problem. Time delay would be the significant cost involved.

PCM appears in two places—the formatting part and the signaling of baseband part. It is a formatting part, because ADC involves sampling, quantization, and yielding binary digits (PAM to PCM) and it is a baseband signaling part because PCM waveforms are used to carry the PCM digits.

The difference between PCM and a PCM waveform is that PCM represents a bit sequence while PCM waveform represents a particular waveform conveyance of that sequence.

11.2 PCM waveform types

When pulse modulation is applied to a binary symbol, the resulting binary waveform is called a PCM waveform. There are several types of PCM waveforms. In telephony applications, those waveforms are often called *line codes.* When pulse modulation is applied to a non-binary symbol, the resulting waveform is called *M-ary* pulse modulation waveform. There are several types of PCM waveforms in practical use.

PCM waveforms fall into the four primary groups: (i) non-return-to-zero (NRZ) PCM; (ii) return-to-zero (RZ) PCM; (iii) phase encoded PCM; and (iv) multilevel binary PCM.

11.2.1 Non-return-to-zero (NRZ)

Non-return-to-zero (NRZ) is the most commonly used PCM waveform in use. Within NRZ PCM, there are the following four subgroups: (i) NRZ-L (L for level) PCM; (ii) NRZ-M (M for mark); and (iii) NRZ-S (S for space).

11.2.1.1 NRZ-L PCM waveform

NRZ-L PCM is the one which is used extensively in real applications. A binary *1* is represented by one voltage level and a binary *0* is represented by another voltage level. There is a change in level whenever the data change from a *1* to a *0* or from a *0* to a *1*. Taking an example of 10110011001 as data stream or a PCM code, Fig. 11.2 shows the traces of the NRZ-L PCM waveform.

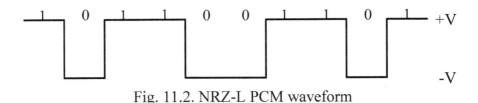

Fig. 11.2. NRZ-L PCM waveform

11.2.1.2 NRZ-M PCM waveform

In NRZ-M PCM waveform, the 1 or mark is represented by a change in level, and the *0* or space, is represented by no change in level. Referred to as differential encoding, NRZ-M PCM is used primarily in magnetic tape recording. Considering the same example of 10110011001, we have Fig, 11.3 showing the resultant NRZ-M PCM waveform.

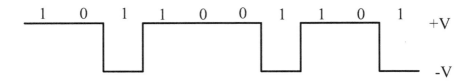

Fig. 11.3. NRZ-M PCM waveform

11.2.1.3 NRZ-S PCM waveform

NRZ-S PCM waveform is the complement of NRZ-M PCM waveform. A *one* is represented by no change in level and a *zero* by a change in level. Our example of 10110011001 PCM code produces the waveform as in Fig. 11.4.

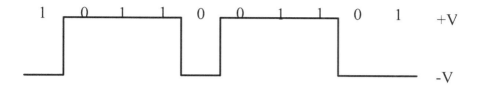

Fig. 11.4. NRZ-S PCM waveform

11.2.2 Return-to-zero (RZ)

The return-to-zero (RZ) waveforms consist of unipolar-RZ, bipolar-RZ, and RZ-AMI (alternate mark inversion). These codes find applications in baseband data transmission and in magnetic recording.

11.2.2.1 Unipolar-RZ PCM waveform

In a unipolar-RZ PCM waveform, a digit 1 is represented by a half bit wide pulse and a *0* is represented by the absence of the pulse. With the same example of 10110011001, Fig. 11.5 illustrates the unipolar-RZ PCM waveform.

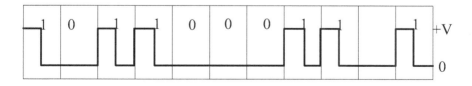

Fig. 11.5. Unipolar-RZ PCM waveform

11.2.2.2 Bipolar-RZ PCM waveform

Bipolar-RZ PCM waveform is constructed by representing *1s* and *0s* by opposite-level pulses that are half bit wide. There is a pulse present in each bit interval. Again, with an example of 10110011001, Fig. 11.6 is the resultant bipolar-RZ PCM waveform.

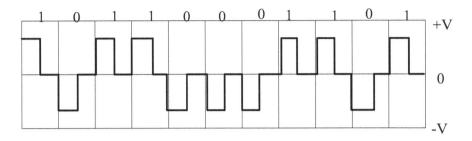

Fig. 11.6. Bipolar-RZ PCM waveform

11.2.2.3 RZ-AMI PCM waveforms

RZ-AMI PCM waveforms, where AMI implies alternate mark inversion, is a signaling scheme used in telephone systems and the *1s* are represented by equal amplitude alternating pulses. The *0s* are represented by the absence of pulses as in Fig. 11.7 for the case of our example of 10110011001.

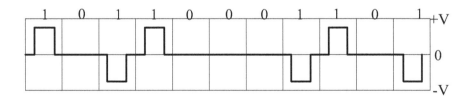

Fig. 11.7. RZ-AMI PCM waveform

11.2.3 Phase encoded

The group of Phase encoded PCM consists of (i) bi-phase-level $(bi - \phi - L)$ or Manchester coding; (ii) bi-phase-mark $(bi - \phi - M)$; (iii) bi-phase-space $(bi - \phi - S)$; and (iv) delay modulation (DM) or Miller coding.

Phase-encoded PCM waveforms are very much used in magnetic recording systems, optical communications systems, and some satellite telemetry links.

11.2.3.1 Bi-phase-level $(bi - \phi - L)$ PCM waveforms

In bi-phase-level $(bi - \phi - L)$ PCM waveforms, a *1* is represented by a half-bit wide pulse positioned during the first half of the bit interval while a *0* is represented by a half-bit wide pulse positioned during the second half of the bit interval. Example in Fig. 11.8 illustrates the $(bi - \phi - L)$.

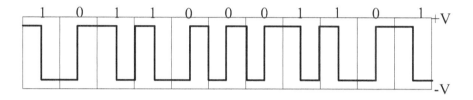

Fig. 11.8. Bi-phase-level $(bi - \phi - L)$ PCM waveforms

11.2.3.2 Bi-phase-mark $(bi - \phi - M)$ PCM waveforms

A transition occurs at the beginning of every bit interval in bi-phase-mark $(bi - \phi - M)$ PCM waveforms. A 1 is represented by a second transition one-half bit wide pulse later. A zero is represented by no second transition as in example shown in Fig. 11.9.

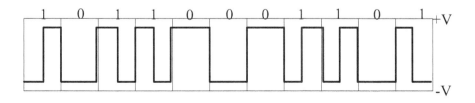

Fig. 11.9. Bi-phase-mark $(bi - \phi - M)$ PCM waveforms

11.2.3.3 Bi-phase-space $(bi - \phi - S)$ PCM waveforms

Bi-phase-space $(bi - \phi - S)$ PCM waveforms is characterized by a transition at the beginning of every bit interval. A *1* is represented by no second transition while a *0* is represented by a second transition one-half bit interval later. An example is depicted in Fig. 11.10.

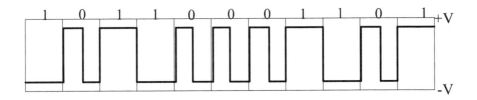

Fig. 11.10. Bi-phase-space $(bi - \phi - S)$ PCM waveforms

11.2.3.4 Delay modulation (DM) or Miller coding PCM waveforms

In Miller coding PCM waveforms, a *1* is represented by a transition at the midpoint of the bit interval while a *0* is represented by no transition, unless it is followed by another zero. In this case, a transition is placed at the end of the bit interval of the first zero. Fig. 11.11 explains the construction of the Miller coding PCM waveforms.

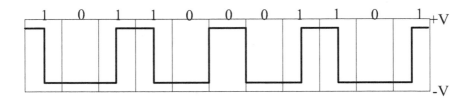

Fig. 11.11. Delay modulation (DM) or Miller coding PCM waveforms

11.2.4 Multilevel Binary

To encode binary data, many binary waveforms use three levels, instead of two. This group has, among others, bipolar-RZ and RZ-AMI. The group also contains formats called di-code and duo-binary.

11.2.4.1 Di-code-NRZ PCM waveforms

The one-to-zero or zero-to-one data transition changes the pulse polarity and without a data transition, the zero level is sent. Fig. 11.12 shows and example of the di-code-NRZ PCM waveforms.

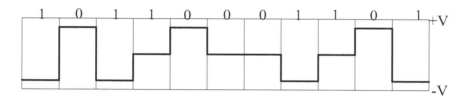

Fig. 11.12. Di-code-NRZ PCM waveforms

11.2.4.2 Di-code-RZ PCM waveforms

Fig. 11.13 shows an example of di-code-RZ PCM waveforms where the one-to-zero or zero-to-one transition produces a half-duration polarity change. Otherwise, a zero level is sent

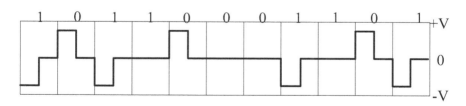

Fig. 11.13. Di-code-RZ PCM waveforms

11.2.4.3 Duo-binary signaling PCM waveforms

Each pulse of the sequence $\{y_k\}$ out of the digital system can be expressed as in Eq. (11.2):

$$y_k = x_k + x_{k-1} \quad\text{---} \quad (11.2)$$

$\{y_k\}$'s are not independent. Each y_k digit carries with it the memory of the prior digit. The estimated $\{y_k\}$ is denoted by $\{\hat{y}_k\}$ and estimated $\{x_k\}$ is denoted by $\{\hat{x}_k\}$.

To understand this scheme, let us see an example that demonstrates duo-binary coding. If we have a sequence $\{x_k\} = 0010110$ and that the first bit is a start up digit which is not a part of the data. Consider Table 11.1 which represents the relationship among $\{x_k\}$, x_{k-1}, and $\{y_k\}$.

Table 11.1. Representation of the relationship among $\{x_k\}$, x_{k-1}, and $\{y_k\}$

Binary digit sequence $\{x_k\}$	0	0	1	0	1	1	0
Polar amplitudes $\{x_{k-1}\}$	-1	-1	+1	-1	+1	+1	-1
Coding rule $y_k = x_k + x_{k-1}$		-2	0	0	0	2	0

The decoding decision rule for the duo-binary signaling PCM waveforms is expressed as follows:

(a) If $\hat{y}_k = 2$, decide that $\hat{x}_k = +1$ (or binary 1)
(b) If $\hat{y}_k = -2$, decide that $\hat{x}_k = -1$ (or binary 0)
(c) If $\hat{y}_k = 0$, decide opposite of the previous decision.

Therefore, the decoded binary sequence $\{\hat{x}_k\}$ will be: 0100110.

To conclude, it is worth to wonder why we do have so many PCM waveforms. In fact the main reason for the large selection relates to the differences in performance that characterize each waveform. In choosing a PCM waveform, the following parameters are worth to examine:

(a) *dc components:* eliminating the dc energy from the signal's power spectrum enables the system to be ac coupled
(b) *self clocking:* symbol or bit synchronization is required for any digital communication system and, hence, the transition provided by Manchester coding, for example, guarantees a self-clocking signal
(c) *error detection:* some schemes, like duo-binary, provide the means for detecting data errors
(d) *bandwidth compression:* some schemes, like multi-level codes, increase the bandwidth efficiency
(e) *differential encoding:* the polarity of differentially encoded data may invert without affecting the data detection.
(f) *noise immunity:* some of the schemes are more immune than others to noise

11.3 Generation and demodulation of PCM

11.3.1 Generation of PCM

Generation of PCM is a complex business. To generate the PCM waveforms, we essentially undergo at least five separate processes: (i) the signal is sampled and converted to PAM; (ii) the PAM samples are quantized; (iii) the quantized PAM samples are encoded based on the alphabet of symbols in a given system; (iv) the supervisory signals are added; and (v) the signal is then sent directly via cable or modulated and transmitted.

Because PCM is highly immune to noise, AM may be used, so that PCM-AM is quite common. Fig. 11.14 presents a schematic diagram of PCM generating system.

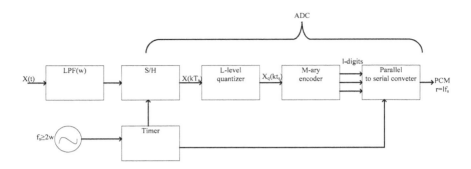

Fig. 11.14. Schematic diagram of PCM generating system

As shown in Fig. 11.14, the analog input waveform $x(t)$ is low-pass filtered and sampled to obtain $x(kT_s)$. A quantizer rounds-off the sample values to the nearest discrete values in a set of L quantum levels. The resulting quantized samples $x_q(kT_s)$ are discrete in time, because of sampling, and discrete in amplitude, because of quantization. The parameters $M, l,$ and L must satisfy the expression in Eq. (11.3) or Eq. (11.4):

$$L = M^l \text{ .... (11.3)}$$

$$l = \log_M L \text{ .... (11.4)}$$

11.3.2 PCM receiver

Consider a PCM receiver with the reconstruction system as shown in Fig. 11.15:

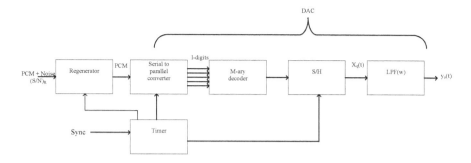

Fig. 11.15. Schematic diagram of PCM **receiver**

The received signal may be contaminated by noise, but regeneration yields a clean and nearly errorless waveform if $(S/N)_R$ is sufficiently large. LPF produces the smoothed output signal $y_D(t)$ which differs from the message $x(t)$ to the extent that the quantized samples differ from the exact sample values $x(kT_s)$. Perfect message reconstruction is, therefore, impossible in PCM, even when random noise has no effect. The ADC operation at the transmitter introduces permanent errors that appear at the receiver as quantization noise in the reconstructed signal.

11.4 Differential PCM

Differential PCM (DPCM) is quite similar to ordinary PCM. However, in this case, each word indicates the difference in amplitude, positive or negative, between the recent sample and the previous sample. The relative value of each sample is indicated and not the absolute value as in normal PCM, It would take fewer bits to indicate the size of the amplitude-change than the absolute amplitude. Therefore, a smaller bandwidth would be required for the transmission.

Nevertheless, the DPCM system has not found wide acceptance. This is because of complications in encoding and decoding process that appears to outweigh any advantages gained.

In DPCM, at each sampling time, say the kth sampling time, the difference between the sample value $m(k)$ at sampling time k and the sample value $m(k-1)$ at time $(k-1)$ can be transmitted. A waveform identical in form to $m(t)$ shall be generated at the receiver by simply cumulatively adding up all these changes.

The special merit is that when these differences are to be transmitted by PCM, the difference $m(k) - m(k-1)$ will be smaller than the sample values themselves. This attracts a fewer levels that will be required to quantize the differences and that means a fewer bits will be needed to encode the levels.

If $\hat{m}(t)$ is the waveform referred to as the approximation to the original message or signal $m(t)$, the difference $m(t) - \hat{m}(t)$ is the precise last change in $m(t)$.

In spite of all the positive attributes that digital or quantized systems have, they indeed face several problems also. The problems with a digital or quantized system include: (i) the fact that the differences are not generally transmitted exactly because of the quantization; (ii) the differences may be larger than the maximum that can be accommodated because of the restricted number of encoding bits we have provided in a particular system; (iii) there might be a large discrepancy between the original signal $m(t)$ and the signal $\hat{m}(t)$ generated at the receiver.

To avoid the above problems, a duplicate of the receiver accumulator is made available at the transmitter so that the same signal $\hat{m}(t)$ is available at the transmitter and that the transmitted signal should convey the difference between $m(t)$ and $\hat{m}(t)$.

Therefore, we add or subtract from $\hat{m}(t)$ a value which is appropriate to bring $\hat{m}(t)$ closer to $m(t)$.

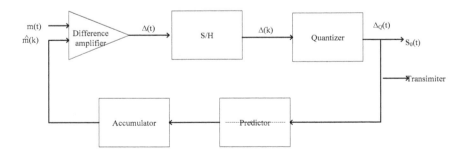

Fig. 11.16. DPCM transmitter

Fig. 11.17. DPCM receiver

Figs. 11.16 and 17, respectively show the DPCM transmitter and receiver. The accumulator at the receiver adds up the received quantized differences $\Delta_Q(k)$. The filter smoothes out the quantization noise. The output of the accumulator is the signal approximation $\hat{m}(k)$ which becomes $\hat{m}(t)$ at the filter output.

At the transmitter, we need to know whether $\hat{m}(t)$ is larger or smaller than $m(t)$, and by how much. We may then determine whether the next difference $\Delta_Q(k)$ needs to be positive or negative and of what amplitude in order to bring $\hat{m}(t)$ as close as possible to $m(t)$. At each sampling time, the transmitter difference amplifier compares $m(t)$ and $\hat{m}(t)$ and the S/H circuit holds the results of that comparison $\Delta(t)$ for the duration of the interval between sampling times. The quantizer generates the signal $S_o(t) = \Delta_Q(t)$ both for the transmission to the receiver and to provide input to the receiver accumulator in the transmitter. The quantized differences will first be encoded at the transmitter and decoded at the receiver.

Chapter Twelve

Delta Modulation

12.1 Introduction

Delta modulation (DM) is a PCM scheme in which the difference signal $\Delta(t)$ is encoded into just a single bit. This technique results into providing for just two possibilities used to increase or decrease the estimated $\hat{m}(t)$.

12.2 Linear Delta modulation

Let us consider a way in which a DM can be assembled as illustrated in Fig. 12.1 with Fig. 12.2 representing a linear DM.

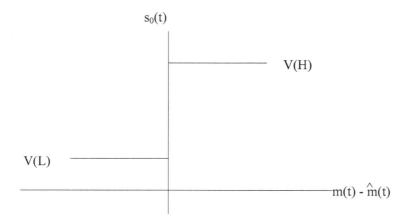

Fig. 12.1. DM waveform

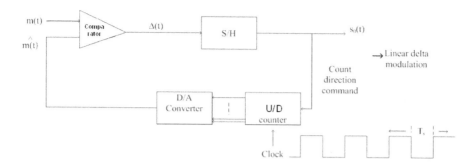

Fig. 12.2. Linear DM

The baseband signal $m(t)$ and its quantized approximation $\hat{m}(t)$ are applied to a comparator as inputs. A comparator simply makes a comparison between inputs. The comparator has one fixed output given by Eq. (12.1) and Eq. (12.2):

$$V(H) \text{ when } m(t) > \hat{m}(t) \quad\text{---} \quad (12.1)$$

and

$$V(L) \text{ when } m(t) < \hat{m}(t) \quad\text{---} \quad (12.2)$$

As $m(t) - \hat{m}(t)$ passes through zero, we have an abrupt transition. In the present case, we need to know only whether $m(t)$ is larger or smaller than $\hat{m}(t)$ and not the magnitude of the difference.

U/D Converter increments or decrements its count by 1 at each active edge of the clock waveform. This means that the count direction, incrementing or decrementing, is determined by the voltage levels at the "count direction command" input to the counter. When the binary input, which is also the transmitted output $S_o(t)$, is at the level $V(H)$, the counter counts-up and when it is at the level $V(L)$, the counter counts down.

The counter serves as the accumulator, since it adds or subtracts increments directed and stores the accumulated result.

The digital output of the counter is converted to the analog quantized approximation $\hat{m}(t)$ by the D/A converter.

Let us consider the waveforms in Fig. 12.3 for the given system assuming that the active clock edge is the falling edge.

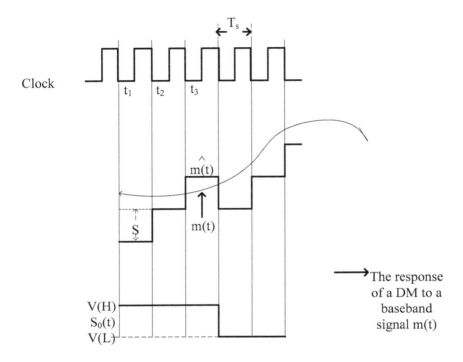

Fig. 12.3. The response of a DM to a baseband signal $m(t)$

For a time period t_1, we have $m(t) > \hat{m}(t)$ which implies that Eq. (12.3) is valid:

$$S_o(t) = V(H) \dotfill (12.3)$$

This means that the counter is incremented and it shows that $\hat{m}(t)$ jumps up by an amount equal to step size S.

At t_2, we find $m(t) > \hat{m}(t)$ still which means that $S_o(t) = V(H)$ and there is another upward jump in $\hat{m}(t)$.

At t_3, we find $m(t) < \hat{m}(t)$ still which means that $S_o(t) = V(L)$, the counter decrements and there is a consequent downward jump in $\hat{m}(t)$ by S.

At a start-up, there may be a brief interval when $\hat{m}(t)$ may be a poor approximation to $m(t)$. This means that there is a large initial discrepancy between $m(t)$ and $\hat{m}(t)$.

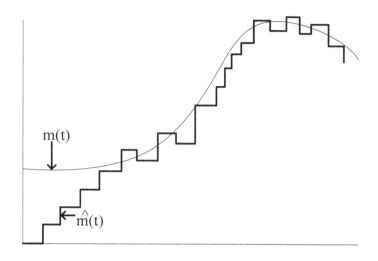

Fig. 12.4. Discrepancy between $m(t)$ and $\hat{m}(t)$ in **DM**

Considering the performance of the system represented by Fig. 12.5 representing slope-overload in the linear DM representing slope-overload error:

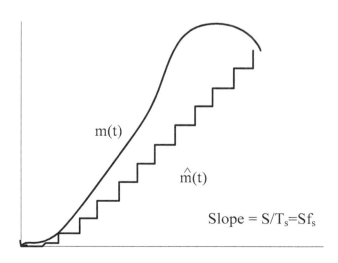

Fig. 12.5. Slope-overload in the linear DM representing slope-overload error

For the case when $m(t)$'s slope is greater than S/T. A signal $m(t)$ over an extended time exhibits a slope which is so large that $\hat{m}(t)$ can not keep up with it. Error $m(t) - \hat{m}(t)$ becomes very large, say much greater than $S/2$. That is $m(t) - \hat{m}(t) >> S/2$.

Linear delta modulation has primitive simplicity, but also the tendency to suffer from severe limitations. Because of that unbalanced traits, linear DM finds almost no applications in real telecommunications systems.

As an example, let us consider a system for speech transmission with a reasonable selection being represented by Eq. (12.4):

$$S = (V_H - V_L)/256 \dashuline{\hspace{3cm}} \quad (12.4)$$

To overcome the overload error, we try to increase the sampling rate above the rate initially selected to satisfy the Nyquist criterion. Since we need only one bit per sample rather than eight, as in PCM, we can increase the sampling rate eightfold before we reach the rate required in PCM. However, experimentally, it has been determined that DM will transmit speech without significant slope overload, given that the DM system is able to transmit a sinusoidal of frequency, $f = 800 Hz$ whose amplitude is the same as the amplitude of the speech waveform.

For amplitude A and frequency f, we have a maximum slope of $2\pi fA$ as it passes through zero. To avoid slope overload, Eq. (12.5) must be satisfied:

$$Sf_s \geq 2\pi fA \dashuline{\hspace{3cm}} \quad (12.5)$$

This means that, $f_s \geq \pi f \dfrac{2A}{S} = 256\pi f \cong 640kHz$

Strange enough, we have set $f = 800 Hz$, but in fact, PCM requires $64 kHz$. This means that DM is not a practical technique for real world telecommunications systems.

12.3 Adaptive delta modulation (ADM)

In adaptive delta modulation (ADM) the step is not kept fixed. When slope overload occurs, the step size becomes progressively larger. This means that $\hat{m}(t)$ is allowed to catch up with $m(t)$ rapidly. Fig. 12. 6 represents a generalized idea of ADM:

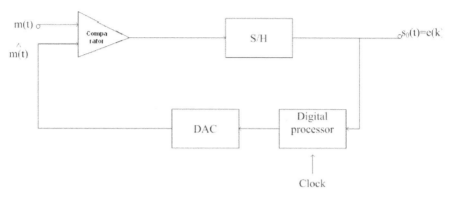

Fig. 12. 6. Generalized idea of ADM

The processor has an accumulator and, at each active edge of the clock waveform, generates a step S which augments or diminishes the accumulator. The step size S is not of fixed size, but it is always a multiple of a basic step S_o.

The output $S_o(t)$ is called $e(k)$ which implies the error $m(t) - \hat{m}(t)$. It is either $V(H)$ or $V(L)$.

$e(k) = +1$ if $m(t) > \hat{m}(t)$ immediately before the kth edge.

$e(k) = -1$ if $m(t) < \hat{m}(t)$ immediately before the kth edge.

At sampling time k, the step size $S(k)$ is given by Eq. (12.6):

$$S(k) = |S(k-1)|e(k) + S_o e(k-1) \quad\text{------------------------------} \quad (12.6)$$

Consider an example in Fig. 12.7 of ADM showing $m(t)$, $\hat{m}(t)$, and $\hat{m}'(t)$.

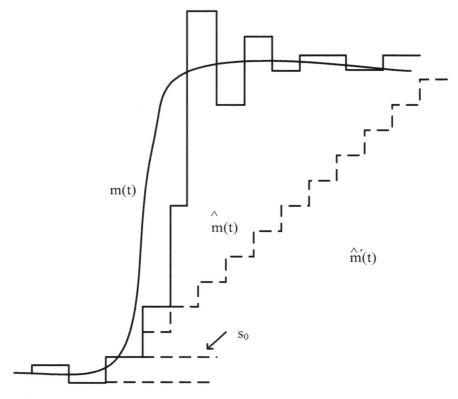

Fig. 12. 7. ADM showing $m(t)$, $\hat{m}(t)$, and $\hat{m}'(t)$

In comparison, ADM can operate at bit rates of $32kbps$ with performance comparable to that obtained when using PCM at $64kbps$. ADM can also operate at $16kbps$ with only a slight degradation of performance. In practice, more than 90% of the step sizes are less than or equal to $2S_o$ and that the probability of a step exceeding $15S_o$ is about 1%.

Chapter Thirteen

Phase Shift Keying (PSK)

13.1 Introduction

Phase shift keying (PSK) technique was developed during the early days of the deep-space program, but it is now widely used in both military and commercial telecommunications systems. It is a digital type of modulation. The general analytical expression for PSK is given in Eq. (13.1):

$$v_i(t) = \sqrt{\frac{2E}{T}} . Cos[\omega_c t + \phi_i(t)] \text{-----------------} (13.1)$$

where $0 \leq t \leq T$ and $i = 1,2,3,...,M$ with $\phi_i(t)$ given by Eq. (13.2):

$$\phi_i(t) = \frac{2\pi i}{M} \text{-----------------------------------} (13.2)$$

13.2 Binary Phase-shift keying (BPSK)

13.2.1 Transmitted BPSK as AM signal

In BPSK, the transmitted signal is a sinusoidal of fixed amplitude. It has one fixed phase when the data is at one level and when the data is at the other level the phase is different by 180^0. If the sinusoid is of amplitude A, the power, P_s, is given by Eq. (13.3):

$$P_s = \frac{1}{2} A^2 \text{-----------------------------------} (13.3)$$

This means that:

$$A = \sqrt{2P_s} \text{---------------------------------------} (13.4)$$

The transmitted signal is either expressed in Eq. (13.5) or Eq. (13.6):

$$v_{BPSK}(t) = \sqrt{2P_s}Cos(\omega_0 t) \text{---} (13.5)$$

or

$$v_{BPSK}(t) = \sqrt{2P_s}Cos(\omega_0 t + \pi) = -\sqrt{2P_s}Cos(\omega_0 t) \text{----------------} (13.6)$$

Data in BPSK is a stream of binary digits, $b(t)$ with voltage levels which are normally taken to be at $+1V$ and $-1V$. When $b(t) = +1V$, it implies logic level 1 and when $b(t) = -1V$ it implies logic level 0.

Without loss of generality, we can safely write Eq. (13.7):

$$v_{BPSK}(t) = b(t)\sqrt{2P_s}Cos(\omega_0 t) \text{-----------------------------------} (13.7)$$

Practically, a BPSK signal is generated by applying the waveform $Cos(\omega_0 t)$, as a carrier, to a balanced modulator and applying the baseband signal $b(t)$ as the modulating waveform. Therefore, BPSK can be thought of as an AM signal.

13.2.2 BPSK reception

The received signal has the form as in Eq. (13.8):

$$v_{BPSK}(t) = b(t)\sqrt{2P_s}Cos(\omega_0 t + \theta) = b(t)\sqrt{2P_s}Cos\omega_0(t + \theta/\omega_0) \text{-------}$$
$$\text{--} (13.8)$$

where θ is a nominally fixed phase shift corresponding to the time delay θ/ω_0 which depends on the length of the path from transmitter to receiver and the phase-shift produced by the amplifiers in the *front-end* of the receiver preceding the demodulator.

The original data $b(t)$ is recovered in the demodulator. Synchronous demodulation is done that requires that there be available at the demodulator the waveform $Cos(\omega_0 t + \theta)$.

A scheme for generating the carrier at the demodulator and for recovering the baseband signal is shown in Fig. 13.1.

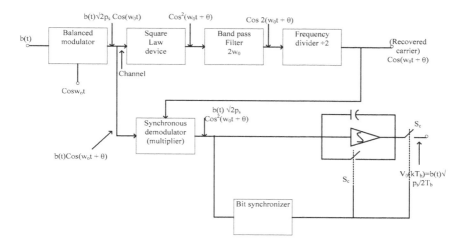

Fig. 13.1.BPSK generation and baseband recovery

The output signal, $v_o(kT_b)$, is expressed as in Eq. (13.9):

$$v_o(kT_b) = b(t)\sqrt{\frac{P_s}{2}}.T_b \quad \text{---} \quad (13.9)$$

The received signal is squared to generate the signal expressed in Eq. (13.10):

$$Cos^2(\omega_0 t + \theta). = \frac{1}{2} + \frac{1}{2}Cos2(\omega_0 t + \theta) \quad \text{-----------------------} \quad 13.10)$$

The bandpass filter, whose passband is centred around $2f_o$, will remove the *dc* components. That means we will have $Cos2(\omega_0 t + \theta)$.

Frequency divider is used to generate the waveform $Cos2(\omega_0 t + \theta)$. Only the waveforms of the signal at the output of the squarer, filter and divider are of interest and not their amplitudes. We can, arbitrarily, take each amplitude to be unity. The carrier having been received is multiplexed with the received signal to generate expression in Eq. (13.11):

$$b(t)\sqrt{2P_s}Cos^2(\omega_0 t+\theta). = b(t)\sqrt{2P_s}\left[\frac{1}{2}+\frac{1}{2}Cos2(\omega_0 t+\theta)\right]$$

.... (13.10)

The resultant in Eq. (13.10) is then applied to an integrator.

A bit synchronizer is able to recognize precisely the moment which corresponds to the end of the time interval allocated to one bit and beginning of the next. At that moment, it closes switch S_c very briefly to discharge the integrator capacitor and leaves the switch S_c open during the entire course of the ensuring bit interval, and so on. This is what we call integrate-and-dump circuit. The output signal is made available by switch S_s which samples the output voltage just prior to dumping the capacitor.

Let the bit interval, T_b, is the duration of an integral number n of cycles of the carrier frequency, f_0. Then, we have Eq. (13.11):

$$n.2\pi = \omega_0 T_b$$ .... (13.11)

Therefore, the output voltage $v_0(kT_b)$ at the end of a bit interval extending from time $(k-1)T_b$ to kT_b is given by Eq. (13.12):

$$v_0(kT_b) = b(kT_b)\sqrt{2P_s}\int_{(k-1)T_b}^{kT_b}\frac{1}{2}dt + b(kT_b)\sqrt{2P_s}\int_{(k-1)T_b}^{kT_b}\frac{1}{2}Cos2(\omega_0 t+\theta)dt$$

.... (13.12)

Therefore,

$$v_0(kT_b) = \sqrt{\frac{P_s}{2}}.T_b$$ .... (13.13)

The system reproduces at the demodulator output the transmitted bit stream $b(t)$. The operation of the bit synchronizer allows to sense each bit independently of every bit. A brief closing of both switches, after each bit has been determined, wipes clean all influence of a preceding bit and allows the receiver to deal exclusively with the present bit.

13.2.3 Spectrum of BPSK

The $b(t)$ waveform is a NRZ binary waveform whose power spectral density is given, for $\left[+\sqrt{P_s}, -\sqrt{P_s}\right]$, by Eq. (13.14)

$$\overline{\left|P(f)\right|}^2 \cong \frac{1}{n}\left\{\left|P_1(f)\right|^2 + \left|P_2(f)\right|^2 + \left|P_3(f)\right|^2 + \ldots + \left|P_n(f)\right|^2\right\}; \ldots (13.14)$$

We have $G_b(f)$ given as in Eq. (13.15):

$$G_b(f) = P_s T_b \left(\frac{Sin\pi f T_b}{\pi f T_b}\right)^2 \ldots (13.15)$$

The power spectral density of the BPSK signal is given by Eq. (13.16):

$$G_{BPSK}(f) = \frac{P_s T_b}{2}\left\{\left(\frac{Sin\pi(f-f_0)T_b}{\pi(f-f_0)T_b}\right)^2 + \left(\frac{Sin\pi(f+f_0)T_b}{\pi(f+f_0)T_b}\right)^2\right\} \ldots (13.16)$$

13.2.4 Interference in BPSK

Inter-channel interference is due to overlap in the spectra of signals which occur in different channels. On the other side, inter-symbol interference (ISI) is a partial overlap of a bit or symbol and its adjacent bits in a single channel. Equalizer at the receiver may alleviate ISI.

13.3 Differential Phase-shift keying (DPSK)

13.3.1 Generation of DPSK signals

In BPSK, to generate the carrier, we start by squaring $b(t)\sqrt{2P_s}Cos(\omega_0 t)$. If the received signal were instead $-b(t)\sqrt{2P_s}Cos(\omega_0 t)$, the recovered carrier would remain as before.

OMAR FAKIH HAMAD

We shall not be able to determine whether the received baseband signal is the transmitted signal $b(t)$ or its negative $-b(t)$.

DPSK and DEPSK are modification of BPSK with the merits that they eliminate the ambiguity about whether the demodulated data is or is not inverted. DPSK avoids the need to provide the synchronous carrier required at the demodulator for detecting BPSK signal.

Let us consider Fig. 13.2 designed for generating a DPSK signal:

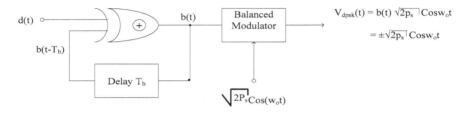

Fig. 13.2. A system for DPSK generation

Table 13.1. Logical values against voltages for $d(t)$, $b(t-T_b)$ and $b(t)$

\multicolumn{2}{c}{$d(t)$}		\multicolumn{2}{c}{$b(t-T_b)$}		\multicolumn{2}{c}{$b(t)$}	
Logical level	voltage	Logical level	voltage	Logical level	voltage
0	-1	0	-1	0	-1
0	-1	1	1	1	1
1	1	0	-1	1	1
1	1	1	1	0	-1

The data stream to be transmitted, $d(t)$, is applied to one input of an X-OR logic gate.

To the other gate input is applied to output of the X-OR gate $b(t)$ delayed by the time T_b allocated to one bit → $b(t-T_b)$→ the second input. The logic waveforms to illustrate the response $b(t)$ to an input $d(t)$ are shown in Fig. 13.3.

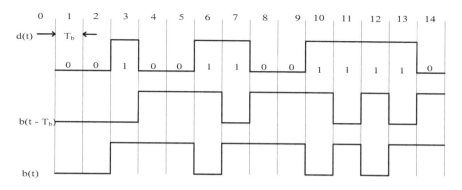

Fig. 13.3. The logic waveforms to illustrate the response *b(t)*
to an input *d(t)*

The waveforms for *d(t)*, *b(t-T$_b$)*, *and b(t)* are consistent with one another. *b(t-T$_b$)* is indeed *b(t)* delayed by one bit time and that $b(t) = d(t) \oplus b(t - T_b)$.

Let during the kth interval. the logic levels of $d(t)$ and $b(t)$ be symbolized as:

$$d(k) \leftrightarrow d(t) \text{ and } b(k) \leftrightarrow b(k)$$

To solve the problem at the first interval, we arbitrarily assume that in the first interval $b(0) = 0$. This $b(t)$ is then applied to a balanced modulator to which is also applied the carrier $\sqrt{2P_s}Cos(\omega_0 t)$. The transmitted signal which is the modulator output is given by Eq. (13.17):

$$v_{DPSK}(t) = b(t)\sqrt{2P_s}Cos(\omega_0 t) = \pm\sqrt{2P_s}Cos(\omega_0 t) \quad\text{............. (13.17)}$$

When $d(t) = 0$, the phase of the carrier does not change at the beginning of the bit interval, while when $d(t) = 1$, there is a phase change of magnitude π.

13.3.2 Recovering the data bit stream from the DPSK

Consider Fig. 13.4 that shows a systematic way on recovering the data bit stream from the DPSK. Here the received signal and the received signal delayed by the bit time T_b are applied to a multiplier. The multiplier output is given by Eq. (13.18): The integrator will suppress the double frequency term.

$$b(t)b(t-T_b)(2P_s)Cos(\omega_0 t + \theta)Cos[\omega_0(t-T_b) + \theta]$$

$$= b(t)b(t-T_b)P_s\left\{Cos\omega_0 T_b + Cos\left[2\omega_0(t-\frac{T_b}{2}) + 2\theta\right]\right\} \ldots\ldots (13.18)$$

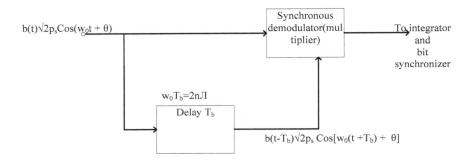

Fig. 13.4. Recovering the data bit stream from the DPSK

We should select $\omega_0 T_b$ so that $\omega_0 T_b = 2n\pi$ with n an integer. This will give $Cos\omega_0 T_b = \pm 1$. This implies that signal output will be as large as possible. The transmitted data bit $d(t)$ can be readily determined from the product $b(t)b(t-T_b)$.

The advantage of the differentially coherent system, DPSK, over the coherent BPSK system is that DPSK avoids the need for complicated circuitry used to generate a local carrier at the receiver. On the other side, the disadvantage of DPSK over PSK is the fact that in a PASK system, an error would be determined. In DPSK a bit determination is made on the basis of the signal received in two successive bit intervals. Therefore, noise in one bit interval may cause errors to two bit determinations. All in all, the error rate in DPSK is greater than in PSK.

13.4 Differentially-Encoded PSK (DEPSK)

The scheme of DEPSK eliminates the need for a device which operates at the carrier frequency and provide a delay of T_b. This means that there is a reduced hardware complication in terms of the DEPSK design.

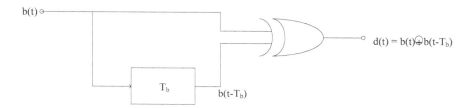

Fig. 13.5. Baseband decoder to obtain d(t) from b(t) in DEPSK

DEPSK transmitter is identical to the transmitter of the DPSK. The recovered signal b(t) is applied directly to one input of an X-OR gate and to the other input b(t-T_b). Depending on whether $b(t) = b(t - Tb)$, or $b(t) = \overline{b(t - Tb)}$, the gate output will be at one or the other of its levels.

In DPSK, bit errors may occur in pairs, but single errors also are possible. In DEPSK, errors always occur in pairs. In DEPSK, a firm definite hard decision is made in each interval about the value of b(t) .

Let us now consider an example in Table 13.2 where *b(k)*, *b(k-1)*, *d(k)*, *b'(k)*, *b'(k-1)*, and *d'(k)* along with their associated errors are computed.

Table 13.2. The computation of *b(k)*, *b(k-1)*, *d(k)*, *b'(k)*, *b'(k-1)*, and *d'(k)* along with the errors

										Time →
b(k)	0	1	1	0	1	1	0	0		One error
b(k-1)		0	1	1	0	1	1	0	0	
d(k) = b(k) ⊕ b(k-1)		1	0	1	1	0	1	0		One error
b'(k)	0	1	1	1	1	1	0	0		

b'(k-1)		0	1	1	1	1	1	0	0	
d'(k) = b'(k) ⊕ b'(k-1)		1	0	0	0	0	1	0		Two errors

In dealing with DEPSK, we conceptually assume error-free signals, b(k), b(k-1), and d(k) = b(k) ⊕ b(k-1). We note that if b(k) has a single error, then, b'(k-1) must also have a single error. The reconstructed waveform d'(k) now has two errors.

13.5 Quaternary PSK (QPSK)

It is a four-phase or quadrature PSK. It is also called quadric-phase shift keying. It is one of the existing multiple phase-shift keying (MPSK). For typical coherent M-ary PSK (MPSK) systems, the signal, $S_i(t)$ can be expressed as in Eq. (13.19).

$$S_i(t) = \sqrt{\frac{2E}{T}} . Cos\left(\omega_0 t - \frac{2\pi i}{M} \right) = \sqrt{2P_s} . Cos\left(\omega_0 t - \frac{2\pi i}{M} \right) \dots (13.19)$$

where $0 \leq t \leq T$ and $i = 1,2,3,...,M$.

In QPSK system, the possible phase-shifts are $+135^o$, $+45^o$, -45°, and -135°. Two bits of information can be indicated instead of one as in the other systems. When a data stream whose bit duration is T_b is to be transmitted by BPSK, the channel bandwidth must be nominally $2f_b$ where Eq. (13.20) holds.

$$f_b = \frac{1}{T_b} \text{ -- (13.20)}$$

QPSK allows bits to be transmitted using half the bandwidth. The description of the QPSK system needs the essential characteristics of type-D flip-flop which is a one bit storage device.

13.5.1 Type-D flip-flop

It is represented in Fig. 13.6. It has a single data input terminal, D, to which a data stream, d(t), is applied.

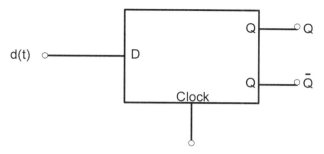

Fig. 13.6. Representation of type-D flip-flop

The Q waveform is the *d(t)* waveform delayed by one bit interval T_b. This implies that once the flip-flop, in response to an active clock edge, has registered a data bit, it will hold that bit until updated by the occurrence of the next succeeding active edge. An example in Fig. 13.7 shows a flip-flop characteristics for the case of d(t) = 00110100110011, where t = 0, 1, 2, 3, ...

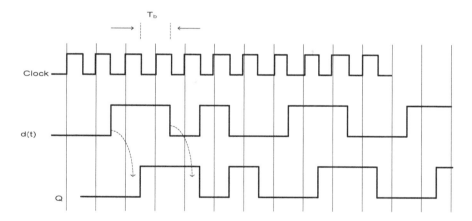

Fig. 13.7. A flip-flop characteristics for *d(t) = 00110100110011*

The delays results from the fact that some time is required for the input data to propagate through the flip-flop to the output Q.

13.5.2 QPSK transmitter

Fig. 13.8 shows a QPSK transmitting mechanism by which a bit stream *b(t)* generates a QPSK signal for transmission.

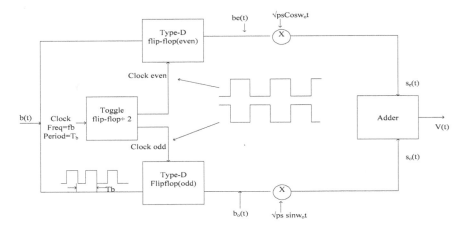

Fig. 13.8. A QPSK transmitter

The toggle flip-flop generates an odd clock waveforms and an even waveforms. These clocks have period $2T_b$, the active edges being separated by the bit time T_b. The bit stream *b(t)* is applied as the data input to both, type-D flip-flops, one driven by the odd and one driven by the even clock waveforms.

The bits in the stream are alternated and held each registered bit for two bit intervals, $2T_b$.

$b_o(t)$, the output of the flip-flop driven by the odd clock, registers bit 1, then bit 3, then bit 5, etc.

$b_e(t)$, holds for times $2T_b$ each, the alternate bits numbered 2, 4, 6, etc

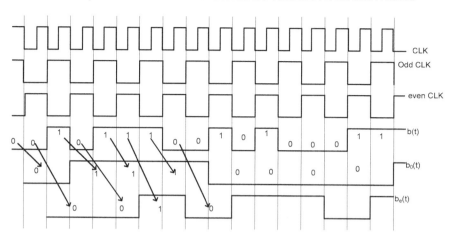

Fig. 13.9. Type-D flip-flop driven by the odd and
the even clock waveforms

The bit stream $b_e(t) = \pm 1 Volt$ is superimposed on a carrier $\sqrt{P_s}.Cos\omega_0 t$ and the bit stream $b_o(t) = \pm 1 Volt$ is superimposed on a carrier $\sqrt{P_s}.Sin\omega_0 t$ by the use of two multipliers, i.e, balanced modulator, to generate two signals $S_e(t)$ and $S_o(t)$.

$S_e(t)$ and $S_o(t)$ are then added to generate the transmitted output signal $v(t)$ which is given by Eq. (13.21).

$$v(t) = \sqrt{P_s}.b_e(t)Cos\,\omega_0 t + \sqrt{P_s}.b_o(t)Sin\omega_0 t \quad\text{.........................} \quad (13.21)$$

The total normalized power of $v(t) = P_s$ which is obtained from Eq. (13.22):

$$v(t) = 2.\frac{P_s}{2}Cos^2\omega_0 t + 2.\frac{P_s}{2}Sin^2\omega_0 t = 2.\frac{P_s}{2}\left[Cos^2\omega_0 t + Sin^2\omega_0 t\right] = P_s$$

$$\text{---} \quad (13.22)$$

In BPSK, a bit stream with bit time T_b multiplies a carrier, the generated signal has a nominal bandwidth $2f_b$. In the waveforms for QPSK $b_o(t)$ corresponds to $\dfrac{1}{2T_b}$ and $b_e(t)$ corresponds to $\dfrac{1}{2T_b}$.

Therefore, both $S_o(t)$ and $S_e(t)$ have nominal bandwidths which are half the bandwidth in BPSK. We have the following:

When $b_o = 1$, the signal $S_0(t) = \sqrt{P_s}\,.Sin\omega_0 t$

When $b_o = -1$, the signal $S_o(t) = -\sqrt{P_s}\,.Sin\omega_0 t$

When $b_e = \pm 1$, the signal $S_e(t) = \pm\sqrt{P_s}\,.Cos\omega_0 t$

Representing the above four signals as phasors, we have a mutual phase quadrature as shown in Fig. 13.10.

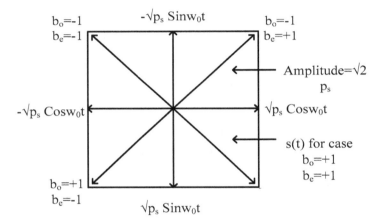

Fig. 13.10. Phasor representation of the mutual phase quadrature

The phasors show the 4 possible output signals which can be generalized in Eq. (13.23);

$$v(t) = S_0(t) + S_e(t) \quad\text{(13.23)}$$

The 4 possible output signals are in phase Quadrature. Either b_o or b_e can change, but both can not change at a time.

13.5.3 A QPSK receiver

Consider a QPSK signal receiver as shown in Fig. 13.11. We have $S(t)$ given by Eq. (13.24).

$$S(t) = \sqrt{P_s}.b_e(t)Cos\omega_0 t + \sqrt{P_s}.b_o(t)Sin\omega_0 t \ldots\ldots\ldots (13.24)$$

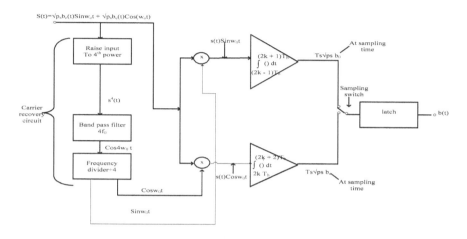

Fig. 13.11. A QPSK receiver

For synchronous detection, it is necessary to locally generate the carriers $Cos\omega_0 t$ and $Sin\omega_0 t$. The incoming signal is applied to two synchronous detectors/demodulators consisting of a multiplier (balanced modulator) followed by an integrator. One demodulator uses the carrier $Cos\omega_0 t$ and the other the carrier $Sin\omega_0 t$.

A bit synchronizer is required for the following reasons: (i) to establish the beginnings and ends of each bit interval; and (ii) to operate the sampling switch. Samples are taken alternatively from one and the other integrator output at the end of each bit time T_b and these samples are held in the latch for the bit time T_b. Therefore, the latch output is the recovered bit stream $b(t)$,

13.6 Differential QPSK (DQPSK)

DQPSK is the mechanism used by QPSK receiver to regenerate the local carrier is a source of phase ambiguity. The carrier may be 180° out of phase with the carrier at the transmitter and as a result the demodulator signals may be complementary to the transmitted signal.

A means for generating a DQPSK signal is shown in Fig. 13.12.

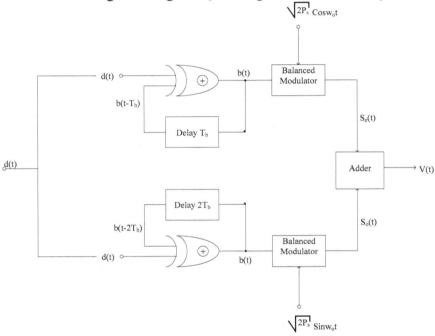

Fig. 13.12. DQPSK signal generator

13.7 M-ary PSK (MPSK)

13.7.1 MPSK Principles

We can recall that, in BPSK, we transmit each bit individually. Depending on whether b(t) is "0" or "1", we transmit one or another of a sinusoid for the bit time T_b, the sinusoids differing in phase by

$\frac{2\pi}{2} = 180^o$. In QPSK we lump together two bits. Depending on which of the four two-bits words develops, we transmit one or another of four sinusoids of duration $2T_b$, the sinusoids differing in phase by amount $\frac{2\pi}{4} = 90^o$. This has not been the end, it is a good news that we can extend the system.

If we lump together N bits and have N-bit symbols, where NT_b satisfies Eq. (13.25):

$$Possible\ symbols = 2^N = M \quad\text{------------------------------} \quad (13.25)$$

If the symbols are represented by sinusoids of duration $NT_b = T_s$, differing from one another by the phase $\frac{2\pi}{M}$, we have a new mechanism called M-ary PSK (MPSK).

The waveforms used to identify the symbols are expressed in Eq. (13.26):

$$v_m(t) = \sqrt{2P_s}\, Cos(\omega_0 t + \phi_m) \quad\text{----------------------} \quad (13.26)$$

where m =0, 1, 2, ..., M-1 and the symbol phase angle, ϕ_m, is expressed by Eq. (13.27):

$$\phi_m = \left((2m+1)\frac{\pi}{M}\right) \quad\text{-----------------------------} \quad (13.27)$$

This implies that:

$$v_m(t) = \left(\sqrt{2P_s}\,Cos\phi_m\right)Cos\,\omega_0 t - \left(\sqrt{2P_s}\,Sin\phi_m\right)Sin\,\omega_0 t \quad\text{----------} \quad (13.28)$$

If $p_e(t) = \sqrt{2P_s}\,Cos\phi_m$ and $p_o(t) = \sqrt{2P_s}\,Sin\phi_m$, then we have Eq. (13.29):

$$v_m(t) = p_e Cos\,\omega_0 t - p_o Sin\,\omega_0 t \quad\text{-------------------------} \quad (13.29)$$

It must be clear that p_e and p_o can change every $T_s = NT_b$ and can be any of M possible values. In addition, ϕ_m, p_e, and p_o are random processes.

Now, the power spectral densities for $G_e(f)$ and $G_o(f)$ are given in Eqs. (13.30) and (13.31):

$$G_e(f) = \frac{\overline{|P_e(f)|^2}}{T_s} = 2P_sT_s\overline{Cos^2\phi_m}\left(\frac{Sin\pi fT_s}{\pi fT_s}\right)^2 \quad \text{-----------------} \quad (13.30)$$

and

$$G_o(f) = \frac{\overline{|P_o(f)|^2}}{T_s} = 2P_sT_s\overline{Sin^2\phi_m}\left(\frac{Sin\pi fT_s}{\pi fT_s}\right)^2 \quad \text{-----------------} \quad (13.31)$$

Since ϕ_m is uniformly distributed, we have Eq. (13.32):

$$\overline{Cos^2\phi_m} = \overline{Sin^2\phi_m} = \frac{1}{2} \quad \text{-----------------------------------} \quad (13.32)$$

Therefore, $G_e(f)$ and $G_o(f)$ are equally expressed as in Eq. (13.33):

$$G_e(f) = G_o(f) = P_sT_s\left(\frac{Sin\pi fT_s}{\pi fT_s}\right)^2 \quad \text{------------------------------} \quad (13.33)$$

When signals with spectral density given as in Eq. (13.33) are multiplied by a carrier, the resultant spectrum is centered at the carrier frequency and extends nominally over a bandwidth, B, given in Eq. (13.34).

$$B = \frac{2}{T_s} = 2f_s = \frac{2f_b}{N} \quad \text{--} \quad (13.34)$$

This means that increasing the number of bits, N, per symbol, the bandwidth becomes progressively smaller.

13.7.2 M-ary transmitter

In M-ary coding, the hardware is only of incidental concern and the M-ary transmitter and receiver combination that will be described here is somewhat of specific nature.

The transmitter system in this case is given in Fig. 13.13 where the bit stream is applied to a serial-to-parallel converter.

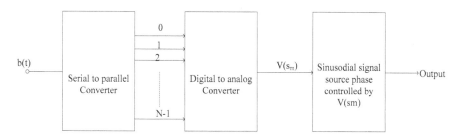

Fig. 13.13. M-ary transmitter system

The N-bits are then presented in parallel. The converter output is applied to a DAC to generate an output voltage which assumes one of $2^N = M$ different values in a one-to-one correspondence to the M possible symbols applied to its input. The DAC output is a voltage $v(S_m)$ which depends on the symbol S_m where $m = 0, 1, 2, ..., M-1$.

13.7.3 M-ary receiver

In M-ary, the receiver is similar to the non-offset QPSK receiver. The recovery system requires to raise the received signal to the Mth power, filter to extract the Mf_o component and then divide by M. Nevertheless, a synchronizer is also needed.

The receiver system is shown in Fig. 13.14 where it has been elaborated that there may or may not be a need to regenerate the bit stream. The idea of transmitting information one bit at a time by a bit stream b(t) arises when we have a system, like BPSK which can handle only one bit at a time.

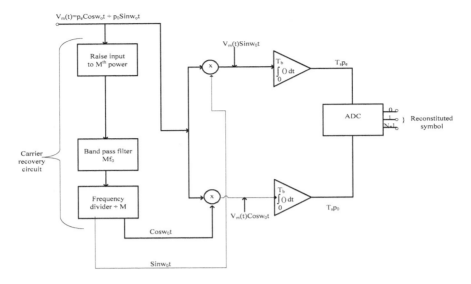

Fig. 13.14. M-ary receiver system

If the system handles M-bit symbols, the data may originate as M-bit words. This implies that there is a need of serial-to-parallel converter. The commonly existing current systems have $M=16$. This means that Bandwidth, B, is given by Eq. (13.35):

$$B = \frac{2f_b}{4} = \frac{f_b}{2}$$ -- (13.35)

This equals to f_b for the case of QPSK.

In conclusion, it must be clear that PSK systems transmit information through signal phase and not through signal amplitude. Therefore, the systems have great merits in situations where the received signal varies in amplitude.

Chapter Fourteen

Frequency Shift keying (FSK)

14.1 Introduction

The general analytic expression for FSK modulation is given in Eq. (14.1).

$$S_i(t) = \sqrt{\frac{2E}{T}} Cos(\omega_i t + \phi) \text{---} (14.1)$$

where $0 \leq t \leq$ and $i = 1,2,3,...,M$,and the frequency term ω_i has M discrete values, and the phase term ϕ is an arbitrary constant.

Consider the FSK waveform sketch in Fig. 14.1 which illustrates the typical frequency changes at the symbols transitions.

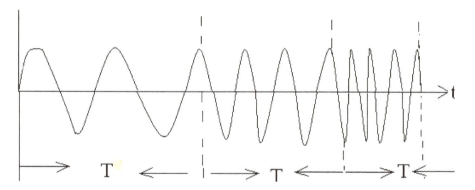

Fig. 14.1. FSK waveforms for the case of $M=3$

At the symbol transitions, there is a gentle shift from one frequency to another. This special class of FSK is called CPFSK (Continuous phase FSK). Conceptually, FSK is represented as in Figs. 14.2 and 14.3.

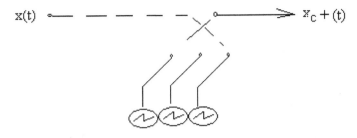

$x(t)$ $x_c + (t)$

Fig. 14.2. Conceptual representation of FSK

$x_c(t)$ Frequency modulator

Fig. 14.3. Conceptual representation of FSK with frequency modulator

The digital signal x(t) controls a switch that selects the modulated frequency from bank of M oscillators.

14.2 Binary frequency shift keying (BFSK)

In BFSK, the binary data waveform d(t) generates a binary signal expresses in Eq. (14.2):

$$v_{BFSK}(t) = \sqrt{2P_s} Cos\left[\omega_0 t + d(t)\Omega t\right] \dotfill (14.2)$$

$d(t) = \pm 1$ corresponds to the logic level 1 or 0 of the data waveform. The transmitted signal is of amplitude $\sqrt{2P_s}$, and is either expressed in Eq. (14.3) or Eq. (14.4).

$$v_{BFSK}(t) = S_H(t) = \sqrt{2P_s} Cos(\omega_0 + \Omega)t \dotfill (14.3)$$

or

$$v_{BFSK}(t) = S_L(t) = \sqrt{2P_s} Cos(\omega_0 - \Omega)t \dotfill (14.4)$$

Therefore, the transmitted signal has an angular frequency $\omega_0 + \Omega$ or $\omega_0 - \Omega$ where Ω is a constant offset from the nominal carrier frequency ω_0. We call: $\omega_0 + \Omega = \omega_H$ the higher frequency and $\omega_0 - \Omega = \omega_L$ the lower frequency

To generate the BFSK signal, two modulators are used; one with carrier ω_H and one with carrier ω_L

14.2.1 The BFSK signal generation

Fig. 14.4 shows the manner in which the BFSK signal is generated:

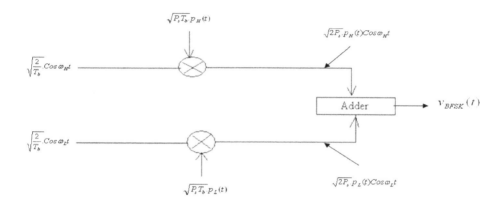

Fig. 14.4. BFSK signal generation

The voltage values of $p_H(t)$ and $p_L(t)$ are related to the voltage values of d(t) as in Table 14.1.

Table 14.1. Relationship between voltages $p_H(t)$ and $p_L(t)$ against voltage values of d(t)

d (t)	$p_H(t)$	$p_L(t)$
+1 V	+1V	0V

-	0V	+1V
1V		

From Table 14.1, hence, when d(t) changes from +1V to -1V, p_H changes from 1 to 0 and p_L changes from 0 to 1. At any time, either p_H or p_L is 1, but not both. Thus, the generated signal is either at angular frequency ω_H or at ω_L.

14.2.2 BFSK spectrum

Considering the variables p_H and p_L, the BFSK signal is given in Eq. (14.5):

$$v_{BFSK}(t) = \sqrt{2P_s}\, p_H Cos(\omega_H t + \theta_H) + \sqrt{2P_s}\, p_L Cos(\omega_L t + \theta_L) \ldots\ldots\ldots \tag{14.5}$$

Each of the terms above looks like the signal $\sqrt{2P_s}b(t)Cos\omega_o t$ that has been encountered in BPSK in which the spectrum is known. In the BPSK case, b(t) is bipolar that alternates between ± 1. In the BFSK case, p_H and p_L are unipolar of either +1 or 0.

Rewriting p_H and p_L as the sum of a constant and a bipolar variable, as in Eq. (14.6) and Eq. (14.7):

$$p_H(t) = \frac{1}{2} + \frac{1}{2}p'_H(t) \ldots\ldots\ldots\ldots \tag{14.6}$$

$$p_L(t) = \frac{1}{2} + \frac{1}{2}p'_L(t) \ldots\ldots\ldots\ldots \tag{14.6}$$

where

p'_H and p'_L are bipolar alternating between +1 and -1.
p'_H and p'_L are complementary which means that when p'_H is +1, p'_L is -1 and vice versa.

Thus, we have Eq. (14.7):

$$v_{BFSK}(t) =$$

$$\sqrt{\frac{P_s}{2}}Cos(\omega_H t + \theta_H) + \sqrt{\frac{P_s}{2}}Cos(\omega_L t + \theta_L) + \sqrt{\frac{P_s}{2}}p'_H Cos(\omega_H t + \theta_H) + \sqrt{\frac{P_s}{2}}p'_L Cos(\omega_L t + \theta_L)$$

-- (14.7)

The first two terms in Eq. (14.7) produce a power spectral density consisting of two impulses; at f_H and f_L. The last two terms produce the spectrum of two BPSK signals; one centered about f_H and one about f_L. The individual power spectral density patterns of the last two terms lokk like here in the figure below for the case $f_H - f_L = 2f_b$

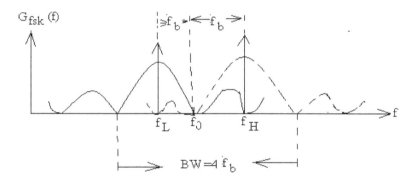

Fig. 14.5. BFSK power spectral density patterns

There is an overlapping between the 2 parts of the spectra, but it is not large and the levels can be distinguished of the binary waveform d(t). Hence, the bandwidth of BFSK is given by Eq. (14.8):

$$BW_{BFSK} = 4f_b$$ -- (14.8)

This means that the bandwidth is twice the bandwidth of BPSK. It is, hence, to be noted that one way to obtain the power spectral density of a waveform is to first obtain the autocorrelation function $R(\tau)$. The power spectral density is the Fourier transform of $R(\tau)$, but $R(\tau)$ is given by Eq. (14.9):

$$R(\tau) = E[v(t)v(t+\tau)] \quad\quad\quad\quad\quad\quad\quad (14.9)$$

We, therefore, first find $R(\tau)$ for the BFSK signal and then, if $\theta_H = \theta_L$, we find $R(\tau)$ and the power spectral density.

14.2.3 BFSK signal receiver

A receiver system in Fig. 14.6 is a typically used system to demodulate a BFSK signal:

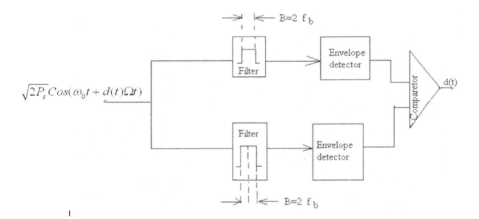

Fig. 14.6. BFSK signal receiver

The signal is applied to two bandpass filters—one with centre frequency at f_H and the other with centre frequency at f_L. The assumption is made that Eq. (14.10) is valid:

$$f_H - f_L = 2\left(\frac{\Omega}{2\pi}\right) = 2f_b \quad\quad\quad\quad\quad (14.10)$$

One filter will pass nearly all the energy in the transmission at f_H and the other will perform similarly for the transmission at f_L.

The envelope detector outputs are compared and a binary output is generated which is at one level or the other depending on which input is larger. Practical systems use a bit synchronizer and an integrator, and sampling is done only once at the end of each time interval T_b.

14.3 Comparison of BFSK and BPSK

From Eq. (14.11), we have:

$$v_{BFSK}(t) = \sqrt{2P_s}\,Cos(\omega_0 t + d(t)\Omega t) \dots\dots\dots\dots\dots (14.11)$$

This implies that Eq. (14.12) is true:

$$v_{BFSK}(t) = \sqrt{2P_s}\,Cos\Omega t Cos\,\omega_0 t - \sqrt{2P_s}\,d(t)Sin\Omega t Sin\omega_0 t \dots (14.12)$$

The second term in Eq. (14.12) looks like BPSK signal with a carrier $Sin\omega_0 t$ multiplied by a data bit d(t) which changes the carrier phase.

In BFSK case, the carrier is not of fixed amplitude. It is shaped by the factor $Sin\Omega t$. Also, the presence of Quadrature reference term $Cos\Omega t Cos\omega_0 t$ is noted. This contains no information, but it carries energy causing the energy in the information bearing term being diminished.

For orthogonal BFSK, each term has the same energy. Hence, the information bearing term contains only one-half of the total transmitted energy.

14.4 M-ary FSK

M-ary FSK (MFSK) is an obvious extension of BFSK. An M-ary FSK communication system is illustrated in Figs. 14.7 and 14.8 with transmitter and receiver sides.

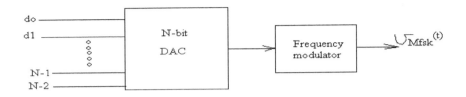

Fig. 14.7. MFSK signal transmitter

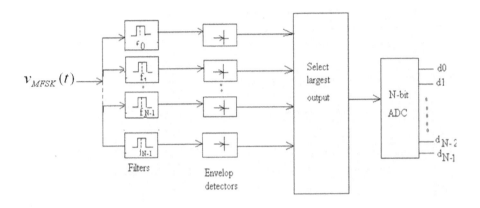

Fig. 14.8. MFSK signal receiver

At the transmitter an N-bit symbol is presented each T_s to an N-bit DAC and the DAC output is applied to a frequency modulator. For the duration T_s, the transmitted signal is of frequency f_0 or f_1 or $f_2 \ldots f_{M-1}$ with $M = 2^N$. The envelope detectors apply their outputs to a device which determines which of the detector indications is the largest and transmits that envelope output to an N-bit ADC.

The commonly employed arrangement simply provides that the carrier frequency be successive even harmonics of the symbol frequency $f_s = \dfrac{1}{T_s}$. This implies that $f_0 = kf_s$; $f_1 = (k2)f_s$; $f_2 = (k+4)f_s$; an so on. To pass ad an M-ary FSK, the required spectral range is: $B = 2Mf_s$, but $f_s = \dfrac{f_b}{N}$ and $M = 2^N$.

Therefore, B is given by Eq. (14.13):

$$B = 2.2^N \frac{f_b}{N} = 2^{N+1} \frac{f_b}{N} \quad\quad\quad\quad\quad\quad\quad\quad\quad\quad (14.13)$$

Hence, M-ary FSK requires a considerably increased bandwidth in comparison with M-ary PSK, but $P_{E(MFSK)}$ decreases as M increases while $P_{E(MPSK)}$ increases with M.

Chapter Fifteen

Quadrature Amplitude Modulation (QAM)

15.1 Introduction

QAM, also called Quadrature-carrier AM, achieves twice the modulation speed of binary ASK (amplitude-shift keying). The functional blocks of a binary QAM transmitter with a polar binary input at rate r_b is as shown in Fig. 15.1.

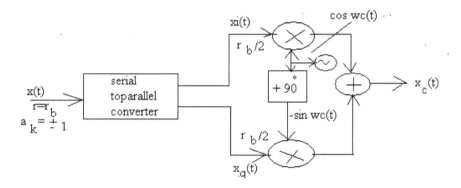

Fig. 15.1. Binary QAM transmitter

The serial-to-parallel converter divides the input into two streams consisting of alternate bits at a rate, r, given by Eq. (15.1)

$$r = \frac{r_b}{2}$$ --- (15.1)

The two modulating signals, $x_i(t)$ and $x_q(t)$, are given by Eq. (15.2) and Eq. (15.3):

$$x_i(t) = \sum_k a_{2k} p_D(t - kD)$$ -- (15.2)

and

$$x_q(t) = \sum_k a_{2k+1} p_D(t - kD)$$ --- (15.3)

where $D = \dfrac{1}{r} = 2T_b$ and $a_k = \pm 1$

The peak modulating values are $x_i = x_q = \pm 1$ during an arbitrary interval. $kD < t < (k+1)D$. This information is conveyed in Fig. 15.2 where there is binary QAM signal representation.

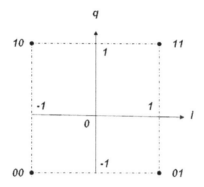

Fig. 15.2. Binary QAM signal representation

Summing the modulated carrier finally yields the QAM signal in the form of Eq. (15.4).

$$x_c(t) = A_c \left[x_i(t) Cos(\omega_c t + \theta) - x_q(t) Sin(\omega_c t + \theta) \right]$$ ---------------- (15.4)

The i and q components are independent but they have the same pulse shape and the same statistical values. That is: $m_a = 0$ and $\sigma^2_a = 1$.

The transmission bandwidth is, thus, reduced to $B_T = \dfrac{r_b}{2}$.

15.2 QAM as a logical extension of QPSK

It can be considered as so because QAM also consists of two independently amplitude-modulated carriers in Quadrature. Each block of k-bits can be split into two $\frac{k}{2}$ bit blocks which use $\frac{k}{2}$ bit DAC to provide the required modulating voltages for the carriers. At the receiver, each of the tow signals is independently detected using matched filters.

QAM can, hence, be thought as the combination of ASK and PSK or just APK. Also, QAM signifies two dimensions ASK which can be modeled as QASK. A canonical QAM modulator is shown in Fig. 15.3.

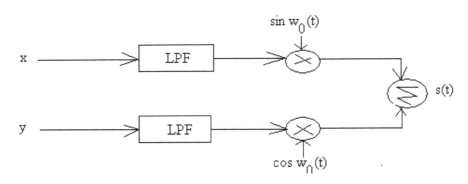

Fig. 15.3. Canonical QAM modulator

Signals are sent in pairs (x, y). The signals are transmitted over separate channels and independently perturbed by noise variables (n_x, n_y), each with *zero* mean and variance N.

15.3 Why do we use QAM (QASK)?

In BPSK, QPSK, and MPSK we transmit, in any symbol interval, one signal or another which are distinguished from one another in *phase,* but are all of the same *amplitude.* To improve the noise immunity of a system, the signal vectors are allowed to differ, not only in their phase, but also in amplitude. This signifies amplitude and phase shift keying system.

Thus QAPSK is about QASK which involves direct balanced modulation of carriers in quadrature. That is, $Cos\omega_0 t$ and $Sin\omega_0 t$. The accepted abbreviation is simply QASK.

Taking an example of QASK system for better understanding, let us suppose that we want to transmit a symbol for every 4-bit. That implies that, we are going to need $2^4 = 16$ different possible symbols. This will enable us to have 16 distinguishable signals that can be able to be generated. We can keep each signal in the geometry such that each signal point is equally distant from its nearest neighbors. The geometrical representation of the 16 signals in a QASK system is as shown in Fig. 15.4.

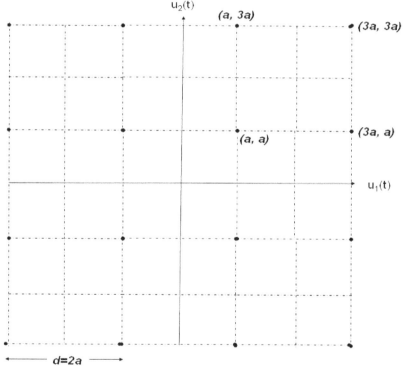

Fig. 15.4. Geometrical representation of the 16 signals in a QASK system

To simplify the hardware design of the system, the points are placed symmetrically about the origin of the signal space. The distance, d, is chosen in such a way that $d = 2a = \sqrt{0.4E_s}$. Therefore, the energy per signal is also kept near a minimum. If we assume that all the 16 signals are equally likely, we can determine the average energy associated with a signal.

From the 4 signals in the first quadrant; the average normalized energy of a signal is given in Eq. (15.5).

$$E_s = \frac{1}{4}\left[(a^2 + a^2) + (9a^2 + a^2) + (a^2 + 9a^2) + (9a^2 + 9a^2)\right]\dots (15.5)$$

$$= \frac{1}{4}\left[40a^2\right] = 10a^2$$

Therefore, $a = \sqrt{0.1E_s}$

Since $d = 2a$, this implies that, $d = 2\sqrt{0.1E_s} = \sqrt{0.4E_s}$

Since each symbol represents 4 bits, the normalized symbol energy is $E_s = 4E_b$ where E_b is the normalized bit energy. Therefore, since $a = \sqrt{0.1E_s}$, we get Eq. (15.6) and Eq. (15.7).

$$a = \sqrt{0.1x4E_b} = \sqrt{0.4E_b} \quad\text{(15.6)}$$

and hence,

$$d = 2\sqrt{0.4E_b} \quad\text{(15.7)}$$

In the case of QPSK signals, $d = 2\sqrt{E_b}$. This means that the distance for QASK is significantly less than the distance between adjacent QPSK signals. However, the distance is greater for 16-QASK than for 16-MPSK where Eq. (15.8) is valid.

$$d = \sqrt{16E_b Sin^2 \frac{\pi}{16}} = 2\sqrt{0.15E_b} \quad\text{(15.8)}$$

A 16 QASK has a lower error rate than 16 MPSK, but a higher error rate than QPSK. From the geometric representation of a typical signal, Eq. (15.9) can be visualized.

$$v_{QASK} = k_1 a u_1(t) + k_2 a u_2(t) \quad \text{(15.9)}$$

where k_1 and k_2 are each equal to ± 1 or ± 3. Since we have Eq. (15.10) and Eq. (15.11):

$$u_1(t) = \sqrt{\frac{2}{T_s}} Cos\omega_o t \quad \text{(15.10)}$$

and

$$u_2(t) = \sqrt{\frac{2}{T_s}} Sin\omega_o t \quad \text{(15.11)}$$

with $a = \sqrt{0.1E_s}$, we get Eq. (15.12):

$$v_{QASK} = k_1 \sqrt{0.2 \frac{E_s}{T_s}}.Cos\omega_o t + k_2 \sqrt{0.2 \frac{E_s}{T_s}}.Sin\omega_o t \quad \text{(15.12)}$$

It is known that $\dfrac{E_s}{T_s} = P_s$. Therefore, we have Eq. (15.13):

$$v_{QASK} = k_1 \sqrt{0.2P_s}.Cos\omega_o t + k_2 \sqrt{0.2P_s}.Sin\omega_o t \quad \text{(15.13)}$$

15.4 QASK signal generation

A generator of a QASK signal for 4-bit symbol is shown in Fig. 15.5 where the four-bit symbol $b_{k+3}, b_{k+2}, b_{k+1}, b_k$ is stored in the 4-bit register made up of 4 flip-flops. A new symbol is presented once per interval $T_s = 4T_b$.

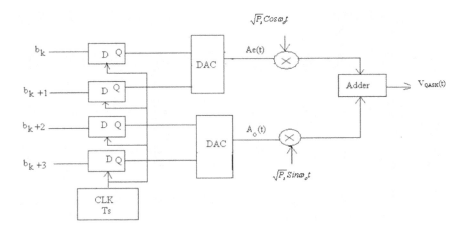

Fig. 15.5. QASK signal generator

Two bits are presented to one DAC and two bits to a second DAC. The converter output $A_e(t)$ modulates the balanced modulator whose input carrier is the even function $\sqrt{P_s}Cos\omega_o t$ while $A_o(t)$ modulates the modulator with off-function carrier $\sqrt{P_s}Sin\omega_o t$. Therefore, the transmitted signal is given in Eq. (15.14):

$$v_{QASK}(t) = A_e(t)\sqrt{P_s}.Cos\omega_o t + A_o(t)\sqrt{P_s}.Sin\omega_o t \quad\text{(15.14)}$$

Comparing the above Eq. (15.14) with the Eq. (15.15):

$$v_{QASK} = k_1\sqrt{0.2\,P_s}.Cos\,\omega_o t + k_2\sqrt{0.2\,P_s}.Sin\,\omega_o t \quad\text{(15.15)}$$

we get Eq. (15.16):

$$A_e,\ A_o = \pm\sqrt{0.2} \text{ or } \pm 3\sqrt{0.2} \quad\text{(15.16)}$$

Also, since all 4 values of A_e and A_o are equally likely, we verify that Eq. (15.17) is valid and, hence, Eq. 15.18).

$$\overline{A_e^2} = \overline{A_o^2} = \frac{1}{2}\left[(\pm\sqrt{0.2})^2 + (\pm 3\sqrt{0.2})^2\right] \quad\text{(15.17)}$$

$$= \frac{1}{2}\left[0.2 + 3^2 x 0.2\right] = \frac{1}{2}.2 = 1$$

Therefore,

$$\overline{A_e^2} = \overline{A_o^2} = 1 \quad \text{...} \quad (15.18)$$

This implies that each of the quadrature terms conveys on the average, one half of the average total transmitted power. The bandwidth of the QASK signal is given by Eq. (15.19):

$$B = 2\frac{f_b}{N} \quad \text{...} \quad (15.19)$$

which is same as in M-ary PSK.

15.5 QASK receiver

The QASK receiver is similar to the QPSK receiver and it is shown in Fig. 15.6.

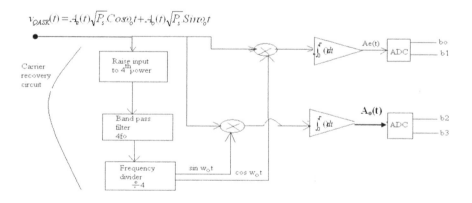

Fig. 15.6. QASK signal receiver

Since A_e and A_o are not of fixed values, we have Eq. (15.20):

$$v^4_{QASK}(t) = P_s^2\left(A_e(t)Cos\omega_o t + A_o(t)Sin\omega_o t\right)^4 \quad \text{...................} \quad (15.20)$$

If we reject all the terms not at the frequency $4f_0$, then we will have Eq. (15.21):

$$\frac{v^4_{QASK}(t)}{P_s^2} = \left[\frac{A_e^4(t) + A_o^4(t) - 6A_e^2(t)A_o^2(t)}{8}\right]Cos4\omega_o t + \left[\frac{A_e(t)A_o(t)[A_e^2(t) - A_o^2(t)]}{2}\right]Sin4\omega_o t$$

-- (15.21)

The average value of the coefficient of $Cos4\omega_o t$ is not zero while the average value of the coefficient of $Sin4\omega_o t$ is zero. This implies that a signal at frequency $4f_o$ will be recovered. Two balanced modulators with two integrators recover the signals $A_e(t)$ and $A_o(t)$.

Chapter Sixteen

Digital Multiplexing Techniques and Hierarchies

16.1 Introduction

When several communication channels are needed between the same two points, significant economies may be realized by sending all, or as many as possible, the messages on one transmission facility. A process is called multiplexing and the reverse is what we call demultiplexing.

By definition, **m**ultiplexing is the sending of a number of separate signals together, over the same cable or bearer, simultaneously and without interference. Fig. 16.1 shows a schematic diagram of multiplexing and demultiplexing scheme where N signals are combined and sent over a common facility during multiplexing process. At the demultiplexing stage, N separate individual signals are recovered.

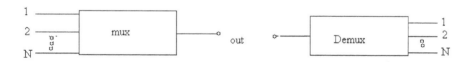

Fig. 16.1. Schematic diagram of Multiplexing and Demultiplexing scheme

There are 2 major kinds of multiplexing techniques—time-division multiplexing (TDM); and frequency-division multiplexing (FDM).

16.2 Digital multiplexing

The same techniques applied to analog signal multiplexing could be applied to waveforms representing digital signals. Digital multiplexing is based on the principle of interleaving symbols from two or more digital signals. The signals to be multiplexed may come from digital data sources or from analog sources that have been digitally encoded.

Dealing with binary multiplexers leads to the accommodation of M-ary signals when necessary by appropriate symbol conversion. Digital Multiplexing is similar to TDM, but digital multiplexing is free from the constraints of periodic sampling and from the waveform preservation.

16.3 Multiplexers and hierarchies

A binary multiplexer (MUX) merges input bits from different sources into one signal for transmission via a digital telecommunications system. The multiplexed signal consists of source digits interleaved bit-by-bit or in clusters of bits—words or characters. For successful demultiplexing at the destination, a carefully constructed multiplexed signal is required with a constant bit rate.

Therefore, a MUX usually must perform four functional operations:

(i)   establish the smallest time interval frame with at least one bit from every input,

(ii) assign a number of unique bit slots, within the frame, to each input,

(iii) insert control bits for frame identification and synchroniza- tion, and

(iv) make allowance for any variations of the input bit rates.

Bit rate variation poses the most annoying design problem in prac- tice.

The mentioned requirements lead to the presence of three major categories of multiplexers:

16.4 Major categories of multiplexers

16.4.1 Synchronous multiplexers

We know that we have a synchronous multiplexer when a master clock governs all the sources. This is characterized by the fact that (a) bit-rate variations are eliminated; (b) the highest efficiency of throughput is attained; and (c) there is complexity for distributing the master-clock signal.

16.4.2 Asynchronous multiplexers

In addition to synchronous multiplexers, we have asynchronous multiplexers which are used for digital data sources that operate in a start/stop mode.

16.4.3 Quasi-synchronous multiplexers

Quasi-synchronous multiplexers are used when the input bit rates have the same nominal values but varying. The input bit rates might be varying within the bounds. These multiplexers, arranged in a hierarchy of increasing bit rates, constitute the building blocks of interconnected digital telecommunications systems.

16.5 Multiplexers hierarchies

For digital telecommunications, two practical slight different multiplexing patterns have been adopted: These are the AT & T hierarchy in North America and Japan and the CCIT hierarchy in Europe. The later is the one adopted by International Telecommunications Union (ITU) and International Telegraph and Telephone Consultative Committee (ITTCC). Both, AT & T and CCIT hierarchies are based on a 64 kbps voice PCM. They have the same structural layout as in Fig. 16.2.

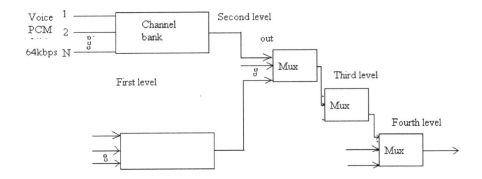

Fig. 16.2. AT&T and CCIT source structural layout

The multiplexing hierarchy in Fig. 16.2 for digital telecommunications has the third level only intended for multiplexing purposes, where as the other three levels are for point-to-point transmission as well as multiplexing. Table 16.1 shows the parameters of AT&T and CCIT hierarchies in practical use in telecommunications systems.

Table 16.1. The parameters of AT&T and CCIT hierarchies:

Hierarchical level	AT&T		CCIT	
	Number of inputs	Output rate	Number of inputs	Output rate
First level	24	1.544 Mbps	30	2.048 Mbps
Second level	4	6.312 Mbps	4	8.448 Mbps
Third level	7	44.736 Mbps	4	34.368 Mbps
Fourth level	6	274.176 Mbps	4	139.264 Mbps

It can be observed that the output bit rate at a given level exceeds the sum of the input bit rates. The surplus allows for control bits and additional stuff bits, needed to yield a steady output rate. This means that, we have very low bandwidth efficiency.

We now consider an **example where** the fourth level of the AT&T multiplexing system transmits 24x4x7x6 = 4032 voice PCM signals with the transmission bandwidth required, B_T, given as: $B_T = \frac{r_b}{2} \equiv 137\,Mbps$. This implies that, the bandwidth efficiency, η_{BW}, is given by Eq. (16.1):

$$\eta_{BW} = \frac{4032x4\,KHz}{137\,MHz} \equiv 12\% \quad\text{..}(16.1)$$

Comparing with the Jumbogroup in the AT&T FDM hierarchy, where 3600 analog voice signals in $B_T = 17\,MHz$ are multiplexed, the bandwidth efficiency, η_{BW}, is given by Eq. (16.2):

$$\eta_{BW} = \frac{3600x4\,KHz}{17\,MHz} \equiv 85\% \quad\text{...}(16.2)$$

Therefore, digital multiplexing sacrifices analog bandwidth efficiency in exchange for the digital transmission's advantages.

Among the advantages of digital transmission include the hardware cost reduction possible by digital integrated circuits; the power cost reduction possible by regenerative repeaters; and the flexibility made possible by digital multiplexing as the input bit streams at any level can be of any desired mix of digital data and digitally encoded analog signals.

If we consider an illustrative configuration in Fig. 16.3 of the AT&T hierarchy with voice, digital data, visual telephone, and color TV signals combined for transmission on the fourth-level T4 line.

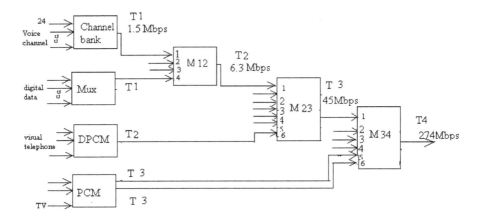

Fig. 16.3. Configuration of the AT & T hierarchy

The first-level T1 signals include PCM voice and multiplexed data. The second-level T2 signals are multiplexed T1 signals along with visual telephone signals encoded as DPCM with $f_s \approx 2MHz$ and 3 bits per word.

PCM encoding of color TV requires a 90-Mbps bit rate. This means that, we must have:

$f_s \approx 10MHz$, 9 bits per word. This means that, two third-level T3 lines are allocated to this signal. M12, M23, and M34 belong to the quasi-synchronous class.

16.6 The Channel Bank

The channel bank is the first-level synchronous multiplexer. T1 voice PCM channel bank synchronous multiplexing of voice PCM requires that the signals be delivered in analog form to the channel bank. A sequential sampling under the control of a local clock generates an analog TDM PAM signals. The signals are transformed by the encoder into TDM PCM with interleaved words. Finally, the processor appends framing and signaling information to produce the output T1 signal. Fig. 16.4 illustrates.

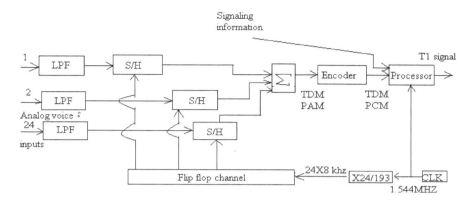

Fig. 16.4. The Channel Bank

With frame bit, FB, the T1 frame structure looks like the one in Fig. 16.5.

FB	Channel 1								Channel 2											Channel 24							
	1	2	3	4	5	6	7	8	1	2	3	4	5	6	7	8	.	.	.	1	2	3	4	5	6	7	8

← 193 bits, 125μsec →

Fig. 16.5. **The T1 Frame structure with a frame bit**

Each frame contains one 8-bit word from each of the 24 input channels plus one bit for framing. This means *one frame = 8x24+1 = 193 bits*. The sampling frequency is 8 kHz with frame duration of 125μsec. Therefore, T1 bit rate, r_b, is given as:

$$r_b = \frac{193bits}{125\mu\sec} = 1.544Mbps$$

Signaling information, that is, dual pulses, busy signals, etc, is incorporated by a method appropriately known as *bit robbing*. In bit robbing, every sixth frame, a signaling bit replaces the least-significant bit of each channel word. Bit-robbing reduces the effective voice PCM word length to $7\frac{5}{6}$. It allows 24 signaling bits every $6x125\mu\sec$, an equivalent signaling rate of $32kbps$.

T1 signals may be either combined at an M12 MUX or transmitted directly over short-haul links for local service up to

80km. The T1 transmission line is a twisted-pair cable with regenerative repeaters every 2km.

Assume that the first-level multiplexer in the CCIT hierarchy is a synchronous voice-PCM channel bank with 30 input signals, output bit rate $r_b = 2.048Mbps$, and no bit-robbing. Let us find the number of framing plus signaling bits per frame.

16.7 Quasi-synchronous Multiplexing

This method of multiplexing becomes necessary when the input bit streams have small variations around some nominal rate. Since the arriving T1 signals were generated by channel banks operating from stable but unsynchronized local clocks, the condition for quasi-synchronous multiplexing exists at an M12 multiplexer. The problems of quasi-synchronous system come in any digital multiplexing system that lacks overall master-clock control.

Quasi-synchronous multiplexers require three essential features: (a) high output bit rate—to accommodate the maximum expected input rates; (b) special buffers called "elastic stores"—for temporary storage of input bits; and (c) bit-buffing—to pad the output stream.

We consider a quasi-synchronous multiplexer presented in Fig. 16.6.

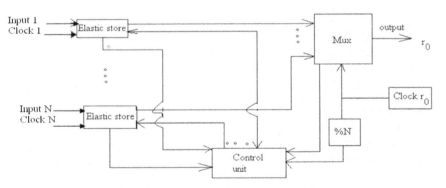

Fig. 16.6. A quasi-synchronous multiplexer

Bit-stuffing together with a synchronous multiplexing make a quasi-synchronous multiplexer. Let that there are N input signals whose rates vary from r_{min} to r_{max} around the nominal value $\bar{r}$. The MUX has a constant output rate $r_o > Nr_{max}$. The control unit receives flip-flop waveforms and returns read signals to each input elastic store.

The unit also supplies the control and stuff bits to the MUX. Let T_f be the frame duration; D be the number of message and stuff bits; and x be the number of control bits. Then, the total number of output bits per frame will be expressed as in Eq. (16.3):

$$T_f r_o = D + x \quad\text{.. (16.3)}$$

Since D is a fixed parameter and the fraction of message and stuff bits depends on the input rates, if all inputs have the same rate, $\bar{r}$, then we will have Eq. (16.4):

$$D = T_f N\bar{r} + Ns \quad\text{.. (16.4)}$$

where s is the average number of stuff bits per channel per frame.

Consequently, if $s = \dfrac{1}{2}$, it implies that each channel gets one stuff bit every other frame under nominal operating conditions. From $T_f r_o = D + x$. This implies that we can have Eq. (16.5):

$$r_o = \frac{D+x}{T_f} = \frac{D+x}{T_f D} T_f N\bar{r}.\frac{D}{D-Ns} \quad\text{............................... (16.5)}$$

Therefore,

$$r_o = N\bar{r}\left(\frac{D+x}{D}\right)\left(\frac{D}{D-Ns}\right) \quad\text{........................... (16.6)}$$

If r_o and s are given, the allowable input rate variations by worst case can be determined. The maximum rate must satisfy the expression in Eq. (16.7):

$$T_f Nr_{max} \leq D \quad\text{... (16.7)}$$

This implies that the upper limit is achieved by a frame with no stuff bits.

At the other extreme, we usually want at most one stuff bit per channel per frame in order to simplify de-stuffing. Therefore, we require that the condition $T_f N r_{min} \geq D - N$ must be satisfied!

From the expressions $T_f r_o = D + x$ and $D = T_f \bar{r} + Ns$, we get Eq. (16.8) and Eq. (16.9):

$$r_{max} \leq \bar{r} + \frac{r_o}{D + x} s \quad\text{---} \quad (16.8)$$

and

$$r_{min} \geq \bar{r} - \frac{r_o}{D + x}(1 - s) \quad\text{-----------------------------} \quad (16.9)$$

This implies that, if we take $s = \dfrac{1}{2}$, we will have Eq. (16.10) and Eq. (16.11)

$$r_{max} \leq \bar{r} + \frac{r_o}{D + x}\left(\frac{1}{2}\right) = \bar{r} + \frac{r_o}{D + x}\left(\frac{1}{2}\right) = \bar{r} + \frac{r_o}{2(D + x)} \quad\text{-------} \quad (16.10)$$

and

$$r_{min} \geq \bar{r} - \frac{r_o}{D + x}\left(\frac{1}{2}\right) \geq \bar{r} - \frac{r_o}{2(D + x)} \quad\text{----------------------------} \quad (16.11)$$

Eq. (16.11) proposes that the system permits equal rate variations around $\bar{r}$.

16.8 The AT&T M12 MUX

M12 MUX of AT&T combines 4 T1 signals with $\bar{r} - 1.555Mbps$. The framing pattern is divided into 4 sub-frames. A complete frame consists of $x = 4x6 = 24$ control bits and $D = 4x6x48 = 1152$ message and stuff bits, where 48 denotes a sequence of 12 interleaved message and stuff bits from each of the 4 inputs. This implies that, we can have r_o computed as in Eq. (16.12)

$$r_o = (4)x(1.544Mbps)x\left(\frac{49}{48}\right)x\left(\frac{288}{288-s}\right) \approx 6.312Mbps \quad (16.12)$$

The value r_o is selected to be $6.312Mbps$ as a multiple of 8kHz and corresponds to $s = \frac{1}{3}$. Therefore, each frame contains $Ns \approx \frac{4}{3}$ stuff bits on the average, for throughput efficiency as computed in Eq. (16.13):

$$\frac{(D-Ns)}{(D+x)} = \frac{N\bar{r}}{r_o} \approx 98\% \quad (16.13)$$

As an exercise, let us consider 24 voice signals arriving at a channel bank already encoded as PCM with $\bar{r} = 64kbps$ .Let that the channel bank is a quasi-synchronous multiplexer whose output frame is divided into 24 sub-frames, each sub-frame containing three control bits and eight message and stuff bits. Let us calculate r_o taking $s = \frac{1}{3}$, and let us determine the throughput efficiency, $\frac{N\bar{r}}{r_o}$.

16.9 Data multiplexers and computer networks

For computer communications, multiplexing differs in two respects: (i) since each computer and data terminal has its own independent clock, it operates in an asynchronous start/stop mode; and (ii) computer and data terminals don't require the nearly instantaneous response needed for two-way voice communications. This implies that extensive buffering is needed and the associated time delay is tolerable.

A third aspect or characteristic of computer communication is relatively low bit rates compared to the megabits per second needed for multiplexed voice PCM. This means that very high internal data transfer speeds within CPU, but this implies that much lower external transmission rates will be required due to the limitations of electromechanical input/output devices.

For example, we can think of the situation when a high speed printer can handle only 50kbps, but multiplexed signals from remote data terminals are typically 10kbps.

The most important question is to know how asynchronous data multiplexers and the CPU are interconnected to form a computer network. Teletypewriters and other keyboard data terminals work with alphanumeric symbols encoded as characters of 7 to 11 bits. These characters are asynchronously transmitted, one at a time, but the bit interval stays fixed throughout all characters.

No, let us consider an example of the 10 bit character format of the ASCII code as in the Fig. 16.7.

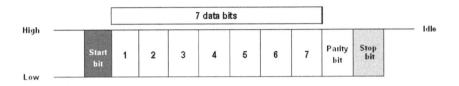

Fig. 16.7. The 10 bit character format of the ASCII

The start and stop bits provide character framing. The parity bit serves as error control, and the remaining seven data bits identify the particular character. Bit rates within the characters of low speed asynchronous terminals range from 75 to 1,200 bps (7200bps for higher speeds). This implies that commercial data multiplexing combines characters from a mix of asynchronous and synchronous terminals. This implies that multiplexed signal is then applied to a modem for transmission over a terminal channel.

A data multiplexer operates in the same basic manner as quasi-synchronous multiplexer, with interleaved characters and character stuffing in place of single bits. This means that frames are comprised of several characters from high speed terminals, but just one character or alternate characters from low speed terminals.

An example of input character buffers is shown in Fig. 16.8.

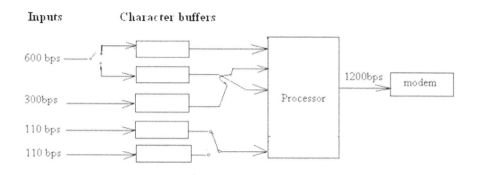

Fig. 16.8. An example of input character buffers

Consider the input rates of 110, 110, 300, and 600bps multiplexed into a 1200bps output, the 80bps surplus allowing for control characters. A sub-frame with two characters from the 600bps input, one from the 300bps input, and alternate characters from the 110bps inputs.

Conventional for non-intelligent data multiplexers has fixed input assignments and transmits dummy characters whenever a particular terminal happens to be idle. Demand-assignment multiplexers have better efficiency as it skips over idle inputs.

Nevertheless, efficient data multiplexing involves significant processing time delays, the delays that would be unacceptable in two-way voice communication, but such delays do not bother machines, and human users of data terminals. Most computer networks operate in a store-and-forward mode that also takes advantage of time-delay tolerance.

Chapter Seventeen

Information Theory and Coding

17.1 Introduction to communication concepts

Telecommunications mainly implies electrical communication, primarily in terms of signals, but the desired information bearing signals are normally corrupted by noise and interference.

Signal theory and analysis is a valuable ideal tool, but it does not grip with the fundamental communication process of information transfer. There is, hence, a need for a broader viewpoint of the communication concepts.

Therefore, a new radical approach is needed. The best of all has been "A Mathematical Theory of Communication" proposed by Claude Shannon back in 1948. This was based on or had connection with the earlier work of Nyquist, Hartley, and Wiener.

Shannon's paper isolated the central task of telecommunications engineering in the question stated here that "Given a message-producing source, not of our choosing, how should the message be represented for reliable transmission over a communication channel with its inherent physical limitations?"

Shannon concentrated on the message information per se, rather than on the signals. This gave birth to "Information Theory" which evolved into a hybrid mathematical discipline and engineering discipline. Information theory deals with three basic concepts: (i) the measure of source information; (ii) the information capacity of a channel; and (iii) coding as a means of utilizing channel capacity for information transfer where coding signifies its broadest sense of message representation, both discrete and analog.

The three concepts, through a series of theorems, came up to the statement that "If the rate of information from a source does not exceed the capacity of a communication channel, then there exists a coding technique such that the information can be transmitted over the channel with an arbitrary small frequency of errors, despite the presence of noise." The three concepts of information theory are presented in Fig. 17.1 where it is about promise for error free transmission on a noisy channel with the help of coding.

Fig. 17.1. Three concepts of information theory

The statement, hence, promises for error free transmission on a noisy channel, with the help of coding. Optimum coding matches the source and channel for maximum reliable information transfer. The coding process involves two distinct encoding and decoding operations. Fig. 17.2 shows coding process that involves two distinct encoding and decoding operations done respectively at the transmitter and at the receiver of a telecommunications system.

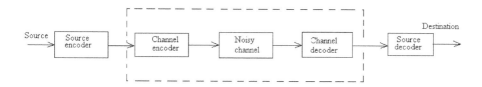

Fig. 17.2. Coding process that involves two distinct encoding/decoding operations

The encoder and decoder units perform the task of error control coding that leads to the reduction of channel noise.

Information theory goes a step further—asserting that optimum channel coding that yield an equivalent noiseless channel with a well defined capacity for information transmission. The source encoder and decoder units then match the source to the equivalent noiseless channel, provided that the source information rate falls within the channel capacity.

17.2 Amount of information

Information, in our discussion, is used as a technical term and should not be confused with *knowledge* or *meaning*. Fig. 17.3 shows information versus knowledge or meaning where there is no amount of information at all in the sense of communication information where Fig. 17.4 implies information versus knowledge or meaning with full amount of information conveyed.

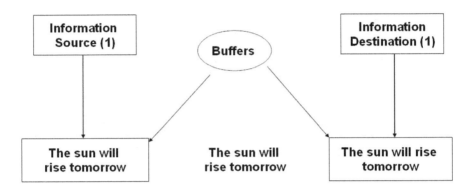

Fig. 17.3. Information versus knowledge or meaning—
no amount of information

The amount of information or measure of information makes the heart of information theory. In telecommunications, information is the commodity produced by the source for transfer to some user at the destination. This implies that the information was previously unavoidable at the destination. Otherwise, the transfer would be zero.

As an example, let use the common example from the literature by supposing that we are planning a trip to a distant city. We need to determine the type of clothes to pack. We call the weather bureau and hear one of the following forecasts: (i) the sun will rise—this conveys no information; (ii) there will be scattered rainstorms—provides information; and (iii) there will be a tornado—more information included as this is an unexpected event.

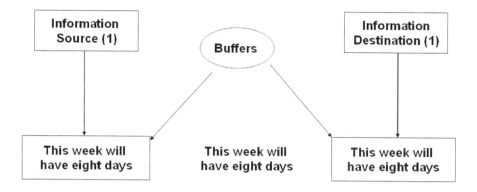

Fig. 17.4. Information versus knowledge or meaning—full amount of information

In principle, the less likely the message, the more information it conveys. Therefore, information measure must be related to *uncertainty* and the uncertainty of the user as to what the message will be corresponds to the freedom of choice by the source in selecting a message.

Whether we prefer the source or user viewpoint, it should be evident that information measure involves the probability of the event or message.

If $P(x_i) = P_i$ is the probability of the event that x_i is selected for transmission, then the amount of information associated with x_i should be some function of P_i.

Shannon defined information measure by the logarithmic function expressed in Eq. (17.1):

$$I_i \underline{\underline{\Delta}} - \log_b P_i = \log_b \left(\frac{1}{P_i} \right) \text{-----------------------------------(17.1)}$$

where b is the logarithmic base, I_i is the self-information of message x_i, and I_i depends only on P_i, irrespective of the actual message content or possible interpretations.

For example, the quoted statement that "the sun will rain tornadoes" would convey lots of information because it is very improbable, despite the lack of substance.

Recalling the definition: $I_i \triangleq -\log_b P_i = \log_b\left(\dfrac{1}{P_i}\right)$, we have several important and meaningful consequences. Some of them are:

(1) $I_i \geq 0$ for $0 \leq P_i \leq 1$ ..... (17.2)

(2) $I_i \to 0$ for $P_i \to 1$ --- (17.3)

(3) $I_i > I_j$ for $P_i < P_j$ --- (17.4)

Therefore, I_i is a non-negative quantity with the properties that:
(a) $I_i = 0$ if $P_i = 1$ which implies that there is no uncertainty
(b) $I_i > I_j$ if $P_i < P_j$ which means that information increases with uncertainty.

Now, let a source produces two successive and independent messages x_i and x_j, with joint probability $P(x_i x_j) = P_i P_j$, then, information,

$I_{ij} = \log_b\left(\dfrac{1}{P_i P_j}\right) = \log_b\left(\dfrac{1}{P_i}\right) + \log_b\left(\dfrac{1}{P_j}\right)$. Hence,

$$I_{ij} = I_i + I_j$$ --- (17.5)

The total information equals the sum of the individual message contributions.

If the logarithmic base b is specified, the unit of information can be determined. Conventionally, b is taken to be equal to 2. Thus, the unit is the *bit* which is the contraction for *binary digit*.

Therefore,

$$I_i = \log_2\left(\dfrac{1}{P_i}\right) bits$$ --- (17.6)

If $P(x_1) = P(x_2)$ the $I_1 = I_2 = \log_2 2 = 1\ bit$ ----------------------- (17.7)

The results in Eq. (17.7) suggest that *1bit* is the amount of information needed to choose between two equally likely alternatives.

It must be carefully distinguished information bits from binary digits per se. A binary digit may convey more or less than one bit of information, depending upon its probability. Therefore, to prevent misinterpretation, *binits* is used for binary digits as message or code elements.

When necessary, base-2 can be converted to natural or common logarithmic using expression in Eq. (17.8):

$$\log_2 v = \frac{\ln v}{\ln 2} = \frac{\log_{10} v}{\log_{10} 2} \quad\text{-- (17.8)}$$

For example if a system has $P_i = \dfrac{1}{10}$, then

$$I_i = \frac{\ln 10}{\ln 2} = \frac{\log_{10} 10}{\log_{10} 2} = \frac{\log_{10} 10}{0.301} = 3.32 \, bits.$$

As an assignment, let us assume equal numbers of the letter grades A, B, C, D, and F are given in a certain course, and we must fine information in bits that one has received when the instructors tells him/her that his/her grade is not F, and how much more information does someone need to determine his/her grade?

17.3 Entropy and information rate

Consider an information source that emits a sequence of symbols selected from an alphabet of M different symbols. Let $X = x_1, x_2, x_3, ...x_M$ represents the entire set of symbols. If we treat each symbol x_i as the message that occurs with probability P_i and conveys the self-information I_i, then: the set of symbol probabilities must satisfy the expression in Eq. (17.9):

$$\sum_{i=1}^{M} P_i = 1 \quad\text{--- (17.9)}$$

The essential assumptions that need to be made for the applicability of Eq. (17.9) include the fact that (i) the source is stationary which means that the probabilities are constant over time; (ii) successful symbols are statistically independent; and (iii) symbols come from the source at an average rate of r symbols per second. The properties define the model of a discrete memory-less source.

Therefore, the amount of information produced by the source during an arbitrary symbol interval is a discrete random variable. Thus, the possible values are $I_1, I_2, I_3,...I_M$.

The expected information per symbol is given by the statistical average called entropy, $H(X)$, defined as in Eq. (17.10):

$$H(X) \underline{\underline{\Delta}} \sum_{i=1}^{M} P_i I_i = \sum_{i=1}^{M} P_i \log_2 \left(\frac{1}{P_i} \right) bits/symbol \dotfill (17.10)$$

The name *entropy* and the notation H were borrowed by Shannon from a similar expression in statistical mechanics. When a sequence of $n >> 1$ symbols are emitted by a source, then the total information to be transferred is about $nH(X)\,bits$. Since the source produces r symbols per second, the time duration of the sequence is $\frac{n}{r}$.

Thus, the information must be transferred at the average rate given in Eq. (17.11):

$$\frac{nH(X)}{\left(n/r \right)} = rH(X)\,bits\,per\,second \dotfill (17.11)$$

Therefore, the source information rate, R, is defined as in Eq. (17.12):

$$R \underline{\underline{\Delta}} rH(X)\,bits/sec. \dotfill (17.12)$$

This is a critical quantity relative to transmission.

Information from any discrete memory-less source can be encoded as binary digits and transmitted over a noiseless channel at the signaling rate given in Eq. (17.13):

$$r_b \geq R\,binits/second \dotfill (17.13)$$

The value of entropy, $H(X)$, for a given source depends upon the symbol probabilities, P_i and the alphabet size, M. $H(X)$ always falls within the limits in Eq. (17.14):

$$0 \le H(X) \le \log_2 M \quad\quad\quad\quad\quad\quad\quad\quad (17.14)$$

which means that $\log_2 M$ occurs when $P_i = \dfrac{1}{M}$. This implies that 0 corresponds to no uncertainty or freedom of choice. All in all, it implies that $P_j = 1$ when $P_i = 0$ for $i \ne j$.

Let us consider the special, but important case of a binary source which implies that $M = 2$. If $P_1 = p$ and $P_2 = 1 - p$. From the definition of the source entropy, then, we have Eq. (17.15)

$$H(X) \underline{\underline{\Delta}} \sum_{i=1}^{M} P_i I_i = \sum_{i=1}^{M} P_i \log_2 \left(\frac{1}{P_i} \right) \quad\quad\quad\quad (17.15)$$

Then, we can have expression as in Eq. (17.16):

$$H(X) = \Omega(p) \underline{\underline{\Delta}} p \log_2 \frac{1}{p} + (1 - p) \log_2 \frac{1}{(1-p)} \quad\quad (17.16)$$

In our discussion, $\Omega(p)$ is called the *horseshoe* function. The plot of $\Omega(p)$ is shown in Fig. 17.5.

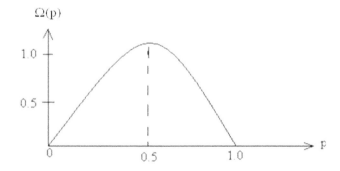

Fig. 17.5. The plot of $\Omega(p)$ - a horseshoe function

A broad maximum is centred at $p = 1 - p = \dfrac{1}{2}$

where $H(X)$ can be computed as in Eq. (17.17):

$$H(X) = \log_2 2 = 1\,\text{bit/symbol} \qquad\qquad (17.17)$$

$H(X)$ decreases monotonically to zero as $p \to 1$ or $1 - p \to 1$

Proving that $0 \leq H(X) \leq \log_2 M$ can be partly done, by separately considering the lower and the upper bounds.

The lower bound is considered by noting that $v \log_2\left(\dfrac{1}{v}\right) \to 0$ as

$v \to 0$ and the upper bound by considering that $H(X) \leq \log_2 M$ which needs an effort of a few more steps. We assume that there is an arbitrary M and we want to prove that the upper bound $H(X) \leq \log_2 M$ is valid.

Let us introduce yet another set of probabilities $Q_1, Q_2, Q_3, ..., Q_M$ to

replace $\log_2 \dfrac{1}{P_i}$ with $\log_2 \dfrac{Q_i}{P_i}$. Subsequently, the conversion from base-2

to natural logarithms gives the quantity expressed in Eq. (17.18):

$$\sum_{i=1}^{M} P_i \log_2\left(\frac{Q_i}{P_i}\right) = \frac{1}{\ln 2} \sum_{i=1}^{M} P_i \ln\left(\frac{Q_i}{P_i}\right) \qquad\qquad (17.18)$$

Recalling the inequality $\ln v \leq v - 1$ and that the inequality holds only if

$v = 1$, let $v = \dfrac{P_i}{Q_i}$ and recalling the equation $\sum_{i=1}^{M} P_i = 1$, we get expres-

sion in Eq. (17.19):

$$\sum_{i=1}^{M} P_i \ln\left(\frac{Q_i}{P_i}\right) \leq \sum_{i=1}^{M} P_i\left(\frac{Q_i}{P_i} - 1\right) = \left(\sum_{i=1}^{M} Q_i\right) - 1 \qquad\qquad (17.19)$$

The obvious condition, $\sum_{i=1}^{M} Q_i = 1$, is imposed. Therefore, we will have

Eq. (17.20):

$$\sum_{i=1}^{M} P_i \log_2 \left(\frac{Q_i}{P_i} \right) \leq 0 \quad\text{--}\quad (17.20)$$

Taking $Q_i = \frac{1}{M}$, we have Eq. (17.21) and, hence, Eq. (17.22):

$$\sum_{i=1}^{M} P_i \log_2 \left(\frac{1}{M} \cdot \frac{1}{P_i} \right) = \sum_{i=1}^{M} P_i \log_2 \left(\frac{1}{P_i M} \right) = \sum_{i=1}^{M} P_i \log_2 \left(\frac{1}{P_i} \right) - \sum_{i=1}^{M} P_i \log_2 M$$

$$\text{--} \quad (17.21)$$

$$= H(X) - \log_2 M \leq 0 \quad\text{------------------------------------}\quad (17.22)$$

This implies that $H(X) \leq \log_2 M$ which was to be proved. The equality holds only in the equally likely case when $P_i = \frac{1}{M}$.

We take an example of a source emitting symbols with $r = 2000$ *symbols per second* that selects from an alphabet of size $M = 4$ with the symbol probabilities and self information as in Table 17.1.

Table 17.1. Symbol probabilities and self information of the
M=4 alphabet

x_i	A	B	C	D
P_i	$\frac{1}{2}$	$\frac{1}{4}$	$\frac{1}{8}$	$\frac{1}{8}$
I_i	1	2	3	3

From the standard definition, we have the source entropy, H(X), given by Eq. (17.23):

$$H(X) = \sum_{i=1}^{4} P_i \log_2 \left(\frac{1}{P_i} \right) \quad\text{--------------------------------}\quad (17.23)$$

$$= \frac{1}{2} x 1 + \frac{1}{4} x 2 + \frac{1}{8} x 3 + \frac{1}{8} x 3$$

$= 1.75 \ bits/symbol \leq \log_2 4 = 2 .$

For the case of the information rate, R, it is given by Eq. (17.24):

$$R = rH(X) \text{...} \quad (17.24)$$

$= 2000x1.75 = 3500 \ bits/second$

This implies that an appropriate coding should make it possible to transmit the source information at the binary signaling rate, $r_b \geq 3500 \ binits/second$.

As an exercise, let a source with $M=3$ symbols with probabilities $P_1 = p$ and $P_2 = P_3$. Show that $H(X) = \Omega(p) + 1 - p$. Then, sketch $H(X)$ versus p .

Again, the entropy is defined as in Eq. (17.25):

$$H(X) = \sum_{i=1}^{M} P_i \log_2 \left(\frac{1}{P_i} \right) \text{...} \quad (17.25)$$

Given that $M=3$ and $P_1 = p$, we can deduce that:

$$P_2 = P_3 = \left(\frac{1-p}{2} \right) \text{..} \quad (17.26)$$

Therefore, $H(X) = \sum_{i=1}^{3} P_i \log_2 \left(\frac{1}{P_i} \right)$

$$= p \log_2 \left(\frac{1}{p} \right) + 2 \frac{(1-p)}{2} \log_2 \left(\frac{2}{1-p} \right)$$

$$= p \log_2 \left(\frac{1}{p} \right) + (1-p) \log_2 \left(\frac{2}{1-p} \right)$$

$$= p \log_2 \left(\frac{1}{p} \right) + (1-p) \log_2 2 - (1-p) \log_2 (1-p)$$

$$= p\log_2\left(\frac{1}{p}\right)+(1-p)+(1-p)\log_2\left(\frac{.1}{1-p}\right)$$

$$= p\log_2\left(\frac{1}{p}\right)+(1-p)\log_2\left(\frac{1}{1-p}\right)+1-p \text{..................} (17.27)$$

However, the term $p\log_2\left(\frac{1}{p}\right)+(1-p)\log_2\left(\frac{1}{1-p}\right)$ corresponds to $\Omega(p)$. Therefore, we will have the Eq. (17.27) simplified to Eq. (17.28):

$$H(X)=\Omega(p)+1-p \text{...} (17.28)$$

The sketch of $H(X)$ versus p can be easily drawn by visually extending that of $\Omega(p)$ and it is left as an exercise for the readers to appreciate.

17.4 Baud rate

If D is the pulse-to-pulse interval, not necessarily equal to the pulse duration, or the time allotted to one symbol, then the signaling rate, r, is define by Eq. (17.29):

$$r\triangleq\frac{1}{D} \text{ symbols/second} = baud \text{.................................} (17.29)$$

In binary, $M=2$, we write $D=T_b$ which corresponds to the bit duration. Then, the bit rate is given as in Eq. (17.30):

$$r=\frac{1}{D}=\frac{1}{T_b} \text{ bits per second (bps)} \text{.................} (17.30)$$

The notations T_b and r_b will be used to identify results that apply only for binary signaling.

Let an amplitude-modulated pulse train expressed by Eq. (17.31):

$$x(t) = \sum_k a_k p(t - kD)$$ .... (17.31)

where a_k is the kth symbol in the message sequence.

The un-modulated pulse $p(t)$ may be rectangular or some other shape, subject to the conditions in Eq. (17.32):

$$p(t) = \begin{cases} 1; t=0 \\ 0; t=\pm D, \pm 2D, \pm 3D, \dots \end{cases}$$ .... (17.32)

The message can be recovered by sampling $x(t)$ periodically at:

$$t = kD\,;\ k = 0, \pm 1, \pm 2, \pm 3, \dots$$ .... (17.33)

since

$$x(kD) = \sum_k a_k p(kD - kD) = a_k$$ .... (17.34)

Therefore, for $\tau \le D$ for $p(t) = 0$ $\forall |t| \ge \dfrac{D}{2}$ then, it is true that Eq. (17.35) is valid.

$$p(t) = \Pi(t/\tau)$$ .... (17.35)

We, therefore, conclude that, baud is the measure, in terms of symbols per second, of the signaling rate, r, defined as in Eq. (17.36):

$$r \triangleq \frac{1}{D}$$ .... (17.36)

17.5 Shannon's theorem and channel capacity

Shannon's theorem is concerned with the rate of transmission of information over a telecommunication channel. Telecommunications channel is an abstraction, intended to encompass all the features and components of the transmission systems which introduce noise or limit the bandwidth.

Shannon's theorem says that, it is possible, in principle, to devise a means whereby a telecommunications system will transmit information with an arbitrarily small probability of error provided that the information rate R is less than or equal to a rate C called the channel capacity. The technique used to approach this end is called *coding*.

17.5.1 Shannon's theorem's statement

"Given a source of M equally likely messages, with $M \gg 1$, which is generating information at a rate R. Given a channel with channel capacity C, then if $R \le C$, there exists a coding technique such that the output of the source may be transmitted over the channel with a probability of error in the received message which may be arbitrarily small."

The significance of the theorem is that, for $R \le C$, transmission may be accomplished without error in even the presence of noise. Shannon's theorem says that noise needs not cause a message to be in error.

17.5.2 Shannon's theorem negative statement

"Given a source of M equally likely messages, with $M \gg 1$, which is generating information at a rate R; then if $R > C$, the probability of error is close to unity for every possible set of M transmitter signals."

17.5.3 Capacity of Gaussian Channel

The Shannon-Hartley theorem is a complementary theorem to Shannon's theorem and applies to a channel in which the noise is Gaussian.

The theorem states that "The channel capacity of a white, band-limited Gaussian channel is $C = B \log_2 (1 + S/N)$ *bits per seconds.*"

Where, in the statement, B is the channel bandwidth, S is the signal power, and N is the total noise within the channel bandwidth, that is given by Eq. (17.37):

$$N = \eta B \quad\quad\quad\quad (17.37)$$

with $\dfrac{\eta}{2}$ the two-sided power spectral density.

From the definitions and statements of Shannon, let X and Y refer to the transmitter output and receiver input, respectively, then we will have:

$H'(X)$ representing the uncertainty or entropy rate of the transmitted signal;

$H'(Y)$ representing the uncertainty or entropy rate of the received signal;

$H'(X/Y)$ representing the uncertainty or conditional entropy of the transmitted signal when the received signal is known; and

$H'(Y/X)$ represents the uncertainty or conditional entropy of the received signal when the transmitted signal is known.

Shannon, then, showed that the rate of transmitting information, R, is given by Eq. (17.38) or Eq. (17.39):

$$R = \left[H'(X) - H'(X/Y) \right] \quad\text{(17.38)}$$

$$= \left[H'(Y) - H'(Y/X) \right] \quad\text{(17.39)}$$

where $H'(X)$, $H'(Y)$, $H'(X/Y)$, and $H'(Y/X)$ are in *bits per second*

17.5.4 Definition of the channel capacity

The channel capacity C is the amount of information correctly transmitted per second and it is given in *bits per second*. In any transmission system, noise is always present and must be taken in to account. Noise leads to reduced error-free capacity of a noisy channel. Therefore, R is equal to the rate of transmitting information $H'(X)$ less the uncertainty of what was sent $H'(X/Y)$, or R is the information received $H'(Y)$ less the uncertainty due to noise $H'(Y/X)$. Shannon calls the uncertainty $H'(X/Y)$ the *equivocation*. Therefore, the maximum channel capacity C is the maximum value of R.

This implies that C is the maximum of $\left[H'(X) - H'(X/Y) \right]$.

For the telecommunications engineers, the value of C for a continuous signal in a noisy channel is useful in many cases.

17.5.5 The Hartley-Shannon law of information

If S is the signal power, N is the noise power, it implies that the total voltage, V, is given by Eq. (17.40):

$$V = \sqrt{S + N}$$ .... (17.40)

with the minimum level, V_{min}, given by Eq. (17.41):

$$V_{min} = \sqrt{N}$$ .... (17.41)

Because of noise, there is an uncertainty and no two levels can be distinguished if they are closer than $\sqrt{N}$

Therefore, we can only use $\sqrt{\dfrac{(S + N)}{N}}$ distinguishable levels and, hence, the information content, H, is represented by Eq. (17.42):

$$H = \log_2 \sqrt{\frac{(S + N)}{N}} \; bits.$$ .... (17.42)

If the time taken to send a pulse is t, then the corresponding bandwidth required is W where t and W are related as in Eq. (17.43):

$$t = \frac{1}{2W}$$ .... (17.43)

This implies that the maximum number of pulse, that is, symbols that can be sent per second, is n, as given by Eq. (17.44):

$$n = \frac{1}{t} = 2W \; pulses \; or \; symbols$$ .... (17.44)

Therefore, the maximum of H' is given by Eq. Eq. (17.45):

$$H' = nH = 2W \log_2 \sqrt{\frac{(S + N)}{N}} \; bits \; per \; second$$ .... (17.45)

It implies that, if the communication capacity is C, then the relation between C, H, and T is given by Eq. (17.46):

$$C = \frac{H}{T} \text{ which is a maximum of } H' \text{............................ (17.46)}$$

Therefore,

$$C = 2W \log_2 \sqrt{\frac{(S+N)}{N}} \text{ bits per second............................ (17.47)}$$

Subsequently,

$$H = CT = WT \log_2 \left(\sqrt{\frac{(S+N)}{N}} \right)^2 \text{ bits............... (17.48)}$$

or

$$H = WT \log_2 \left(1 + \frac{S}{N} \right) \text{ bits............................ (17.49)}$$

Eq. (17.49) is the Hartley-Shannon law of information.

To realize the just concluded law, let us exercise and verify that the communication capacity of a noiseless channel transmitting n discrete message symbols per second will be $C = n\log_2 n$ *bits per second*. Secondly, if we assume a binary symmetric channel, the probabilities of the input messages are $P(x_1) = 0.6$ and $P(x_2) = 0.4$. If $P(y_1/x_1) = 0.8$ and $P(y_2/x_1) = 0.2$, can we determine that the mutual information and channel capacity are, respectively, 0.229 *bits* and 0.279 *bits*.

17.6 Coding, coding efficiency, and error probability

There is an importance of coding information for telecommunications purposes. This has a connection with the channel capacity. There are two ways to relate the characteristics of the source to those of the channel. These are *source coding*, and *channel coding*

17.6.1 Source coding

The principle is to use the minimum number of bits required to convey the necessary information efficiently and subject only to the constraints of a fidelity criterion.

Therefore, the aim is to remove redundancy as far as possible without destroying the content or without destroying the nature of the information.

There are basically two familiar coding techniques which are (i) the Shannon-Fano code and (ii) the Hoffman code.

17.6.1.1 The Shannon-Fano coding technique

The messages are arranged in descending order of their probabilities. They are then grouped into nearly equi-probable groups. To each group is assigned the symbol *1* or *0*. The sub-grouping goes on until they are in pairs. For illustration, consider a source producing eight messages, m_1 to m_8, ($m_1, m_2, m_3, ..., m_8$) with respective probabilities, $P_1, P_2, P_3, ..., P_8$, as given in Table 17.2.

Table 17.2. A source producing messages $m_1, m_2, m_3, ..., m_8$ with probabilities, $P_1, P_2, P_3, ..., P_8$

m	m_1	m_2	m_3	m_4	m_5	m_6	m_7	m_8	$\sum_{i=1}^{8} P_i$
P	0.40	0.20	0.15	0.10	0.06	0.04	0.03	0.02	1.00

The operation for Shannon-Fano coding is explained in the Table 17.3.

Table 17.3. The operation for Shannon-Fano coding for the source in Table 17.2

m	P	Coding sequence		Code	Bits
m_1	0.40	1	1	1 1	2x0.40

m_2	0.20			0			1 0	2x0.20
m_3	0.15	0	1	1			0 1 1	3x0.15
m_4	0.10			0			0 1 0	3x0.10
m_5	0.06		0	1	1		0 0 1 1	4x0.06
m_6	0.04			0			0 0 1 0	4x0.04
m_7	0.03			0	1		0 0 0 1	4x0.03
m_8	0.02			0			0 0 0 0	4x0.02
$\sum_{i=1}^{8} P_i$	1.00							2.55

From the definition of entropy of source, for this case, as in Eq. (17.50):

$$H = \sum_{i=1}^{m} P_i \log_2\left(\frac{1}{P_i}\right) = \sum_{i=1}^{8} P_i \log_2\left(\frac{1}{P_i}\right) \quad\text{..............................(17.50)}$$

Therefore, $H = 2.44$ *bits per symbol*

From the technique in Table 17.3, the number of bits is equal to 2.55. Hence, the coding efficiency is calculated as in Eq. (17.51):

$$Efficiency = \frac{2.44}{2.55} = 95.7\% \quad\text{..............................(17.51)}$$

17.6.1.2 Huffman coding technique

The messages are arranged in a column in descending order of probability and the two lowest probabilities are grouped to form a new probability. The new probability and the remaining probabilities are placed in a new column in descending order as before. The procedure is repeated till the final probability is 1.0. Each of the grouped probabilities forms a junction which is assigned the digit 1 or 0. The coding of any message is obtained by following its horizontal line and the appropriate arrow lines to the final point 1.0.

If we consider the example given earlier in section 17.6.1.1 of the messages $m_1, m_2, m_3, ..., m_8$ with the probabilities $P_1, P_2, P_3, ..., P_8$, the operation for Huffman coding is illustrated in Table 17.4 and the codewords listed in Table 17.5.

Table 17.4. The operation for Huffman coding for the source in Table 17.2

m	P							
m_1	0.40	0.40	0.40	0.40	0.40	0.40	0.60	1.00
m_2	0.20	0.20	0.20	0.20	0.25	0.35	0.40	
m_3	0.15	0.15	0.15	0.15	0.20	0.25		
m_4	0.10	0.10	0.10	0.15	0.15			
m_5	0.06	0.06	0.09	0.10				
m_6	0.04	0.05	0.06					
m_7	0.03	0.04						
m_8	0.02							

Table 17.5. The resultant codewords from the Huffman coding technique in Table 17.4

Message	Codewords	Bits
m_1	0	1x0.40
m_2	1 1 1	3x0.20
m_3	0 1 1	3x0.15
m_4	0 0 1	3x0.10
m_5	0 1 0 1	4x0.06
m_6	0 1 1 0 1	5x0.04
m_7	1 1 1 1 0 1	6x0.03
m_8	0 1 1 1 0 1	6x0.02
Total		2.49 *bits*

As shown in Table 17.5, the code digits at each junction along the path traced give the required code in left-to-right sequence.

From the definition of entropy we have a computation of H in our example given in Eq. (17.52):

$$\text{Entropy, } H = \sum_{i=1}^{m} P_i \log_2\left(\frac{1}{P_i}\right) = \sum_{i=1}^{8} P_i \log_2\left(\frac{1}{P_i}\right) = 2.44 bits \ .....(17.52)$$

Since the entropy of source is 2.44 *bits* and the number of bits from Huffman's techniques is equal to 2.49, we can, hence, obtain the efficiency as in Eq. (17.53):

$$\text{Efficiency} = \frac{2.44}{2.49} = 98\% \ .... (17.53)$$

Finally, as an assignment, let us consider a discrete source that transmits six message symbols with probabilities 0.3, 0.2, 0.2, 0.15, 0.1, and 0.05. We must devise suitable Shannon-Fano and Huffman codes for the messages and determine the average length and efficiency of each code.

17.6.2 Channel coding

In channel coding, redundancy is added to combat the affects of transmission errors that have occurred due to noise. Extra check bits or digits are involved and those are in a coded form. Transmitted digits include information digits and error-correcting digits. Most of the development in coding theory has been taking place in the area of channel coding.

To detect and correct channel errors due to noise, error-correcting codes are designed. They are classified as: Block codes or Convolutional codes.

In block codes, the check bits or digits are used to check the information bits in the block while convolutional codes check the information bits in previous blocks.

Typically, an (n, k) code contains a total of n bits, with k information bits and *(n-k)* check bits. A block code $[C]=[c_1c_2c_3...c_n]$ can be defined by a parity-check matrix H if Eq. (17.54) holds.

$$[H]\oplus[C]^T = 0 \ .... (17.54)$$

where $\oplus$ denotes module-2 arithmetic, and $[C]^T$ is the transpose of the row matrix $[C]$ which means that $[C]^T$ is a column matrix.

If $[H] \oplus [C]^T \neq 0$, it leads to an error in the codeword which can be detected by means of a syndrome $[S]$ where $[S]$ is a column matrix yielding a binary number which indicates the digit-error position.

For example, if $[C']$ is the received codeword, let $[H] \oplus [C']^T = [S]$ and if $[S] = \begin{bmatrix} 0 \\ 0 \\ 1 \end{bmatrix}$, then, there is an error in the first digit. Examples of such codes include the parity-check codes, the Hamming codes, and the Bose-Chaudhuri-Hocquenhem (BCH) code

17.6.2.1 Parity-check code

Parity-check code can only detect a single error in a transmitted message. It checks on the number of ones transmitted at the receiver and decides based on odd parity or even parity operation.

For example, if a single-parity-check digit is added to the stream of message digits so that the number of errors transmitted is always even, we say that the systems work in even-parity-check operation. If a single error occurs in transmission, the number of received ones is odd and an error is at once detected, and a re-transmission of the message is requested.

17.6.2.2 Hamming code

The Hamming code is used for detecting and correcting single errors in a series of received binary digits. A set of digits is divided into a group of k information digits and c check digits. That is, the relation between n, k, and c is as given in Eq. (17.55):

$$n = k + c \quad\quad (17.55)$$

In order to check n positions, the relation in Eq. (17.56) must be valid:

$$2^c - 1 \geq n \quad\quad (17.56)$$

Redundancy or extra check digits are used to check errors and for minimum redundancy, Eq. (17.57) must be satisfied:

$$2^c - 1 = n \quad\text{...}\quad (17.57)$$

The parity-check matrix is easily constructed by using the binary equivalent of each column numbered as a decimal number—left to right—as in the example to follow.

Taking and example of four information digits and three check digits, it means that $n=7$ and $k=4$. This means (7, 4) Hamming codewords with an H matrix of the form in Eq. (17.58):

$$[H] = \begin{bmatrix} 0 & 0 & 0 & 1 & 1 & 1 & 1 \\ 0 & 1 & 1 & 0 & 0 & 1 & 1 \\ 1 & 0 & 1 & 0 & 1 & 0 & 1 \end{bmatrix} \quad\text{.............................}\quad (17.58)$$

And a codeword $[C] = [c_1 c_2 k_1 c_3 k_2 k_3 k_4]$ where k_1, k_2, k_3, and k_4 are the information bits and c_1, c_2, and c_3 are the check bits. It must be noted that, $k_i's$ and $c_i's$ must satisfy the linear equations in Eqs. (17.59):

$$\begin{aligned} c_1 &= k_1 \oplus k_2 \oplus k_4 \\ c_2 &= k_1 \oplus k_3 \oplus k_4 \quad\text{..............................}\quad (17.59) \\ c_3 &= k_2 \oplus k_3 \oplus k_4 \end{aligned}$$

Let, for example, that the codeword transmitted was 0 0 1 0 1 1 0 and it was received in error as 0 0 1 0 0 1 0 where the error is in the fifth position. To determine the syndrome $[S]$ indicating the error position in binary form we use module-2 arithmetic to obtain Eq. (17.60):

$$[H] \oplus [C]^T = \begin{bmatrix} 0 & 0 & 0 & 1 & 1 & 1 & 1 \\ 0 & 1 & 1 & 0 & 0 & 1 & 1 \\ 1 & 0 & 1 & 0 & 1 & 0 & 1 \end{bmatrix} \begin{bmatrix} 0 \\ 0 \\ 1 \\ 0 \\ 0 \\ 1 \\ 0 \end{bmatrix} \text{--------------------- (17.60)}$$

This means that Eq. (17.60) can be further simplified to Eq. (17.61):

$$[H] \oplus [C]^T = \begin{bmatrix} 0 \oplus 0 \oplus 0 \oplus 0 \oplus 0 \oplus 1 \oplus 0 \\ 0 \oplus 0 \oplus 1 \oplus 0 \oplus 0 \oplus 1 \oplus 0 \\ 0 \oplus 0 \oplus 1 \oplus 0 \oplus 0 \oplus 0 \oplus 0 \end{bmatrix} \text{--------------------- (17.61)}$$

Therefore,

$$[S] = \begin{bmatrix} 1 \\ 0 \\ 1 \end{bmatrix}$$ -- (17.62)

The syndrome, as in Eq. (17.62), is $[S] = \begin{bmatrix} 1 & 0 & 1 \end{bmatrix}$, which is the binary code for the decimal number 5, confirming that the error is in the fifth position.

17.6.2.3 Bose-Chaudhuri-Hocquenhem (BCH) code

Starting with an example of (n, k) BCH code in which the first set of k bits are the information bits and the last set of *(n-k)* bits are the check bits, it must be noted that the BCH code is also a cyclic code and, so, codewords can be easily generated using a shift register as an encoder, as in Fig. 17.6.

To decode a BCH code, a syndrome is formed by passing the received stream of bits into a shift register similar to the one at the transmitter. A bit by bit comparison of the check bits reveals if an error has occurred and yields a syndrome which is a record of the error pattern received. This implies that the decoder makes corresponding corrections to the information bits and the corrected information bits are read out by the decoder, as shown in Fig. 17.7.

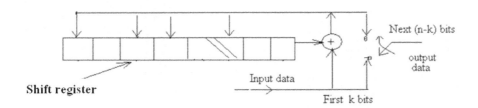

Fig. 17.6. An (n, k) BCH code encoder

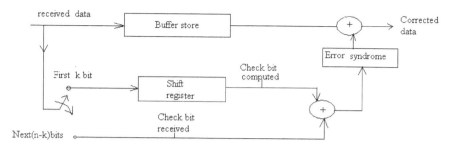

Fig. 17.7. An (n, k) BCH code decoder

BCH codes are widely used for random-error correction. They require the smallest number of check bit for a given reliability. The codes are constructed so that $n = 2^m - 1$ where m is an integer. The codes contain mxe parity check bits which can correct up to e errors and detect up to $2e$ errors.

For example, a (127, 106) BCH code with n=127; m=7; and e=3 can correct up to 3 errors and detect up to 6 errors in any given block.

17.6.2.4 Convolutional codes

A continuous stream of bits is transmitted instead of being coded in blocks. The check bits or digits are used to check overlapping groups of bits known as the *constraint span*. Such codes are generated by an n-stage shift register connected in a predetermined way using feedback.

A convolution code, as shown in Fig. 17.8, can be generated by combining the outputs of the shift register with one or more module-2 adders. The shift register is operated by clock pulse synchronized to the bit stream and, if n is the number of output bits per input, the code rate is $\frac{1}{n}$.

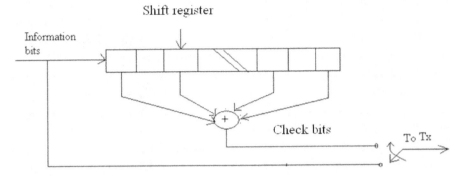

Fig. 17.8. An encoder using n-stage register and a
single module-2 adder

Decoding of the received bit stream does not require any block synchronization and is achieved by checking the received sequence bit-by-bit and making a decision as to whether a *1* or *0* was transmitted. A decoding algorithm which may be employed is called *Viterbi* decoding.

An example of an encoder using an n-stage shift register and a single module-2 adder is shown in Fig. 17.8.

17.6.3 Digital Errors

In the transmission of digital information, errors are bound to occur with a definite probability. This means that in any message consisting of a certain number of digits, the error probability of a received message is of interest.

Let us take for example a message transmitted with a five-digit code, and the probability of a digit being in error is p, then the probability of it being correct is *(1- p)*.

Some cases of the error probability of the message include one-digit error message, two-digit error message, and all-correct message.

17.6.3.1 One-digit error message

For the case of five-digit code, the probability of one error is p and the probability of the remaining four digits being correct is $(1- p)^4$. Therefore, the joint probability of one error digit and four correct digits is $p\ (1-p)^4$. This can occur in $^5C_4$ ways and so the total probability is given by Eq. (17.63):

$$5p(1-p)^4 \approx 5p(1-4p) \quad\text{..}\quad (17.63)$$

since $p<<1$

17.6.3.2 Two-digit error message

The probability of two error digits is p^2 and the probability of the remaining three digits being correct is $(1-p)^3$. Therefore, the joint probability of two error digits and three correct digits is $p^2(1-p)^3$.
This can occur in $^5C_2$ ways and so the total probability is given in Eq. (17.64):

$$10p^2(1-p)^3 \approx 10p^2(1-3p) \quad\text{.......................................}\quad (17.64)$$

since $p<<1$

17.6.3.3 All-correct message

The probability of five digits being correct at the same time is $(1-p)^5$ and this can occur in only one way. Therefore, the total probability is, thus, given by Eq. (17.65):

$$(1-p)^5 \approx (1-5p+10p^2) \quad\text{....................................}\quad (17.65)$$

In conclusion, it can be realized that, since p is usually about 10^{-4}, the probability of two or more error digits in a message is very small. To overall digital errors in transmission, and to maintain these errors at an acceptable level, which is usually called the bit error rate (BER), various error-detecting and error-correcting codes have been devised and some continue to be proposed with slight modifications from the primary ones.
To conclude this chapter, let us recap the concepts on communications information rate. Let the input source to a communication channel (transmitter) be denoted by X and the output source from the communication channel (receiver) be denoted by Y. Fig. 17.9 shows an abstract of the simplified system:

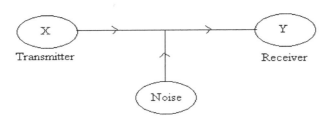

Fig. 17.9. A simplified communication system with a
transmitter, X, and a receiver, Y

For message symbols with probabilities P_i, the entropy of the input source is given by Eq. (17.66):

$$H(X) = -\sum_i P_i \log_2 P_i \quad\text{.. (17.66)}$$

Similarly, for the output source with symbol probabilities P_j, the entropy is given by Eq. (17.67):

$$H(Y) = -\sum_j P_j \log_2 P_j \quad\text{.. (17.67)}$$

The conditional probability that the ith symbol was transmitted when the jth symbol was received is $P(i/j)$ and is related to the joint probability $P(i, j)$ where Eq. (17.68) is valid:

$$P(i, j) = P_j(P(i/j)) = P_i(P(j/i)) \quad\text{............................... (17.68)}$$

Therefore, the conditional entropy per symbol received is $H(X/j)$ given by Eq. (17.69):

$$H(X/j) = -\sum_i P(i/j) \log_2 P(i/j) \quad\text{............................... (17.69)}$$

The average conditional entropy $H(X/Y)$ for all possible received symbols is given by the sum of $H(X/j)$ over all values of j and weighted with probabilities P_j.

This implies that $H(X/Y)$ can be expressed as in Eq. (17.70):

$$H(X/Y) = \sum_j P_j H(X/j) \quad\text{..}\quad (17.70)$$

$$= \sum_j \sum_i P(i/j)\log_2 P(i/j) \quad\text{..................................}\quad (17.71)$$

$$= \sum_i \sum_j P(i,j)\log_2 P(i/j) \quad\text{..................................}\quad (17.72)$$

$$Information = \log_2\left[\frac{a \quad posteriori \quad probability}{a \quad priori \quad probability}\right] \quad\text{..........}\quad (17.73)$$

If the ith symbol was transmitted, then the a priori probability is P_i and if the jth symbol was received, the a posteriori probability is $P(i/j)$

This implies that the information is given as in Eq. (17.74):

$$Information = \log_2\left[\frac{P(i/j)}{P_i}\right] \quad\text{.....................................}\quad (17.74)$$

This is the information transmitted per symbol transmitted and received.

The information transmitted for all transmitted and received symbols is the sum over all joint probabilities $P(i,j)$, that is, $\sum_i \sum_j P(i,j)$

Hence, information, H , is given by Eq. (17.75):

$$H = \sum_i \sum_j P(i,j)\log_2\left[\frac{P(i/j)}{P_i}\right] \quad\text{.............................}\quad (17.75)$$

$$= \sum_i \sum_j P(i,j)\log_2 P(i/j) - \sum_i \sum_j P(i,j)\log_2 P_i \quad\text{...............}\quad (17.76)$$

$$= \sum_i \sum_j P(i,j)\log_2 P(i/j) - \sum_i P_i \log_2 P_i \quad\text{....................}\quad (17.77)$$

since $\sum_i \sum_j P(i,j) = \sum_i P_i \sum_j P_i(j/i) = \sum_i P_i$ .... (17.78)

Therefore, we have H and R given as in Eqs. (17.79) or (17.80) and (17.81), respectively:

$$H = -H(X/Y) + H(X) \quad\quad (17.79)$$

$$= H(X) - H(X/Y) \quad\quad (17.80)$$

and

$$R = H' = H'(X) - H'(X/Y) \quad\quad (17.81)$$

If the discrete source transmits n symbols where x_i is the ith symbol transmitted and y_j is the jth symbol received as illustrated in Fig. 17.10.

From the general expression of $H(X)$, we can have the following set of expressions in Eqs. (17.82) through (17.85);

$$H(X) = -\sum_{i=1}^{n} P(x_i)\log_2 P(x_i) \quad\quad (17.82)$$

$$H(Y) = -\sum_{i=1}^{n} P(y_j)\log_2 P(y_j) \quad\quad (17.83)$$

$$H(X/Y) = -\sum_{i=1}^{n}\sum_{j-1}^{n} P(x_i;y_j)\log_2 P(x_i/y_j) \quad\quad (17.84)$$

and

$$H(Y/X) = -\sum_{i=1}^{n}\sum_{j-1}^{n} P(x_i;y_j)\log_2 P(y_j/x_i) \quad\quad (17.85)$$

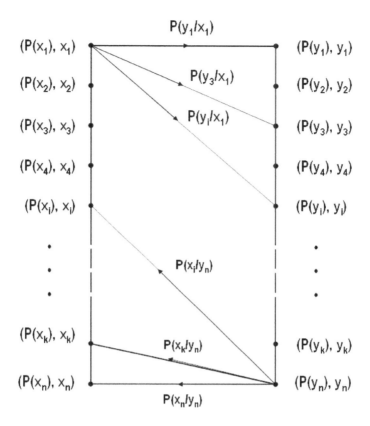

Fig. 17.10. A discrete source with n transmitted symbols, x_i, and n received

This gives the mutual information, $I(X;Y)$, expressed as in Eq.(17.86):

$$I(X;Y) = H(X) - H(X/Y) = H(Y) - H(Y/X) \ \text{...........} \ (17.86)$$

where the joint probability, $P(x_i; y_j)$, is given by Eq. (17.87):

$$P(x_i; y_j) = P(x_i)P(y_j / x_i) = P(y_j)P(x_i / y_j) \ \text{..............} \ (17.87)$$

In the following example, with all the notations in their standard use, we are given $P(x_1) = 0.6$; $P(x_2) = 0.4$; $P(y_1 / x_1) = 0.8$;

$P(y_2/x_1) = 0.2$; $P(y_1/x_2) = 0.2$; and $P(y_2/x_2) = 0.8$, we are to calculate the mutual information, $I(X;Y)$ in that given system.

From the basic definition for the mutual information, we have $I(X;Y)$ given by Eq. (17.88):

$$I(X;Y) = H(X) - H(X/Y) = H(Y) - H(Y/X) \quad \text{(17.88)}$$

we calculate the terms as follows:

(1) $H(X) = -\sum_{i=1}^{2} P(x_i) \log_2 P(x_i)$

$$= -[P(x_1)\log_2 P(x_1) + P(x_2)\log_2 P(x_2)] \quad \text{(17.89)}$$

$$= -[0.6\log_2 0.6 + 0.4\log_2 0.4] = [0.44218 + 0.5288] = 0.971\, bits$$
per symbol

(2) $H(X/Y) = -\sum_{i=1}^{2}\sum_{j-1}^{2} P(x_i;y_j)\log_2 P(x_i/y_j) \quad \text{(17.90)}$

$$= -\begin{bmatrix} P(x_1;y_1)\log_2 P(x_1/y_1) + P(x_1;y_2)\log_2 P(x_1/y_2) \\ + P(x_2;y_1)\log_2 P(x_1/y_2) + P(x_2;y_2)\log_2 P(x_2/y_2) \end{bmatrix}$$

However, from the general expression in Eq. (17.91), $P(x_1;y_1), P(x_1;y_2), P(x_2;y_1)$, and $P(x_2;y_2)$ can be calculated, respectively, from Eqs (17.92) through (17.95):

$$P(x_i;y_j) = P(x_i)P(y_j/x_i) = P(y_j)P(x_i/y_j) \quad \text{(17.91)}$$

$$P(x_1;y_1) = P(x_1)P(y_1/x_1) = 0.6x0.8 = 0.48 \quad \text{(17.92)}$$

$$P(x_1;y_2) = P(x_1)P(y_2/x_1) = 0.6x0.2 = 0.12 \quad \text{(17.93)}$$

$$P(x_2;y_1) = P(x_2)P(y_1/x_2) = 0.4x0.2 = 0.08 \quad \text{(17.94)}$$

and

$$P(x_2;y_2) = P(x_2)P(y_2/x_2) = 0.4x0.8 = 0.32 \quad \text{(17.95)}$$

Substituting the values, we get:

$$H(X/Y) = -0.4[0.2\log_2(0.2) + 0.8\log_2(0.8)]$$
$$- 0.6[0.8\log_2(0.8) + 0.2\log_2(0.2)]$$

$$= \left[0.2\log_2\left(\frac{1}{0.2}\right) + 0.8\log_2\left(\frac{1}{0.8}\right)\right] = \left[0.2\log_2(5) + 0.8\log_2\left(\frac{1}{0.8}\right)\right]$$

$$= 0.4644 + 0.2575 = 0.722 \; bits \; per \; symbol$$

Therefore, $I(X;Y)$ is calculated as:

$$I(X;Y) = H(X) - H(X/Y) = H(Y) - H(Y/X)$$

$$= 0.971 - 0.722 = 0.249 \; bits \; per \; symbol$$

In conclusion, we find out that the mutual information, $I(X;Y)$ in that given system is $0.249 \; bits \; per \; symbol$.

Part III
Multimedia
Telecommunications

Chapter Eighteen

Introduction to Multimedia Telecommunications

18.1 Multimedia telecommunications

Multimedia telecommunications can be defined, in one among many possible definitions, as the transfer (transmission and switching) of computer information in terms of text, graphics, drawings, images audio, video and animation from a point to a point and from multiple points to multiple points. Multimedia, as a field of study, is concerned with the computer controlled integration of text, graphics, drawings, still and moving images or video, animation, audio, and any other media where every type of information can be digitally represented, stored, transmitted and processed.

Multimedia application implies an application which uses a collection of multiple media sources like text, graphics, images, sound or audio, animation and video. The World Wide Web, multimedia authoring like Adobe/Macromedia Director, hypermedia courseware, video-on-demand, interactive TV, computer games, virtual reality, digital video editing and production systems, and multimedia database systems make just a short list among many multimedia applications in existence.

18.2 Multimedia telecommunications system

A telecommunications system capable of processing, storing, buffering, transmitting, switching, and correctly receiving multimedia data and applications is what we call a Multimedia telecommunications system. The said system is essentially characterized by the processing, storage, generation, manipulation, detection and recovery of multimedia information.

We further categorize a multimedia telecommunications system by the possession of the four basic features. These are computer controllability, integrability, digital nature of the information handled and interactivity of the interface.

The challenges for multimedia telecommunications systems are many in this world of growing demand for real time, interactive, shared metadata transportation. For the multimedia telecommunications system to serve the needs of multimedia applications, we have no option but to invest on the system spread over distributed networks' scenario. The challenge of having temporal relationship between data is another task to consider in multimedia telecommunications systems.

The system is also expected to render different data at same time at a continuous manner. The sequencing within the media is another challenge as in some media we need to allow the playing of frames in correct order and with the time frame in a video, for example. Synchronization, which includes inter-media scheduling for media like video and audio, is another challenge to consider while designing multimedia telecommunications system. For us humans to pleasantly watch playback of video and audio or animation and audio, we need to have a system that supports lip synchronization. It is in fact annoying watching an out of lip synchronized film for even a short time clip.

The key issues for multimedia telecommunications systems include the representation and storage of temporal information and strictly maintain the temporal relationships on play back and retrieval of the information. The system must have a process and mechanism to represent the multimedia data in digital form. This means that the multimedia telecommunications system must have the modules for analog to digital conversion, for sampling the signal levels, quantizing the sampled signal levels and encoding into equivalent binary codewords. This means that, there must be a provision for large volume data which involves requirements for wider bandwidths and higher volume storages or robust and reliable compression techniques.

18.3 Features for a multimedia telecommunications system

Since multimedia telecommunications system can not avoid high volumes of data, it is essential that a multimedia telecommunications system has a very high data processing power. This is needed to deal with large data processing and real time delivery of media. As a result, multimedia telecommunications systems need special hardware for the purpose. Equally important, the systems need to have multimedia capable file system to enable the delivery of real-time media like video and audio streaming.

We also need special hardware and software to handle special technology like RAID technology and other modern technologies. Further, the system should represent multimedia data in a format that supports multimedia and should be easy to handle and allow for compression and decompression in real-time.

Efficient and high input/output capability is another prerequisite for the multimedia telecommunications system. This means that input and output to the file subsystem need to be efficient and fast to allow for real-time recording as well as playback of data. To allow access to file system and process data efficiently and quickly, a special operating system is needed in a multimedia telecommunications system. There is a need to support direct transfers to disk, real-time scheduling, fast interrupt processing, input and output streaming and so on.

Large storage units of the order of hundreds GB, and large memory units of the order of GB are required along with a large caches for efficient management of multimedia over multimedia telecommunications systems.

For proper transfer of information over multimedia telecommunications systems, we need a network support of client-server fashion and the systems commonly described as distributed systems. We, indeed, require that multimedia telecommunications systems run under user friendly software tools to easily and correctly handle media, design and develop applications, and deliver the processed media.

18.4 Basic components of a multimedia telecommunications system

For any multimedia telecommunications system to perform its basic functions, there are essential hardware and software components that are required. It is to be clear that these are just the basic ones where the depth of need, the number, and the capacity of those components are variably changing as per the development in technology and the needs put over the multimedia telecommunications systems themselves.

All in all, a multimedia telecommunications system will essentially need capture devices like video camera, video recorder, audio microphone, keyboards, mice, graphics tablets, 3D input devices, tactile sensors, virtual reality devices, and digitizing hardware.

Components like hard disks, CD-ROMs, and DVD-ROMs are among the essential storage devices that are needed in multimedia telecommunications systems. The specifications of each individual component, again, depends on the system under context which directly depends on the amount of data to be processed, the speed and other characteristics of the deployed networks, the urgency and efficiency to be met, and sometimes the real time nature of the application to be served.

Another essential and the backbone of the multimedia telecommunications system is a set of communication networks themselves. A good multimedia telecommunications system is always ready to serve within local networks, within intranets, around the internet, around multimedia or other special high speed telecommunications networks.

The multimedia telecommunications systems rely on the existence of computer controllable data which are the main resource of the system functioning. In that connection, computer systems like multimedia desktop machines, workstations, MPEG/VIDEO/DSP hardware make another essential components of the multimedia telecommunications system.

Finally, since most of the performance of multimedia applications are measured via visual perception, it becomes essential to have display devices like CD-quality speakers, HDTV,SVGA, Hi-Res monitors, and Colour printers. These make an important part of the multimedia telecommunication system, especially at the end user's point of view.

With all those components put in place, the multimedia telecommunications systems can find impressing applications in areas including www, hypermedia courseware, video conferencing, video-on-demand, interactive TV, groupware, home shopping, interactive and kids games, virtual reality, and digital video editing and production systems.

18.5 Data in multimedia telecommunications systems

For a designer wishing to have a working specific purpose multimedia telecommunications system, the consideration of the data to be used is a key step. Knowing at least the input and the format of the data to be processed is necessary. Text and static data, graphics, images, audio and video are the chosen and the common data types practically under consideration in dealing with the functioning of multimedia telecommunications systems. The sources, storage and format characteristics of the five types are here briefly described.

The simplest of all in terms of handling is the text and static data whose source can be keyboards, speech inputs, optical character recognition devices or just data stored on disk. The data is stored and input character by character. For text, the storage is 1 byte per character. For other forms of data like spreadsheet files, we may store the format as text with formatting, but binary encoding may also be used. The format may be either in raw text or formatted text like HTML, Rich Text Format (RTF), Word or a program language source like C, Pascal or C++. The data is not temporal, but **it** may have natural implied sequence like HTML format sequence, sequence of C program statements or C++ program statements. Comparing to other multimedia data in practice, the since of text and static data is very insignificant.

For Graphics, the format arises from the fact that the data is constructed by the composition of primitive objects such as lines, polygons, circles, curves and arcs. The graphics are usually generated by a graphics editor program like Illustrator or automatically by a program like Postscript. Unlike images, graphics are usually editable or revisable. Devices like keyboards—for text and cursor control, mouse, trackball or graphics tablet are the commonly available graphics input devices. OpenGL, PHIGS, and GKS are among the existing graphics standards in use. Graphics files usually store the primitive assembly and they do not take up a very high storage overhead like images, for example.

Images or still pictures, on the other hand, which when uncompressed, are represented as a bitmap, that is, a grid of pixels. The common inputs could be digitally scanned photographs or pictures or direct from a digital camera. The inputs may also be generated by programs, the way it is done with graphics or animation programs. The storage of the image is done at 1 bit per pixel for black and white, 8 bits per pixel for the case pf grey scale or colour map or 24 bits per pixel for the true colour case. In terms of the size, we can imagine that a 512x512 grey scale image can take up to 250 Kb, a 512x512 of 24 bit image with no compression can take about 750 Kb. The overhead increases even higher with image size of modern high digital camera of mega-pixels of 10 and above. It climbs higher to about 29Mb when uncompressed. Here we can see the importance of data compression which is commonly applied.

As well discussed in Part I of this book, audio signals are continuous analog signals which is a representation of the real world information. The inputs for the audio data can be by microphones and then digitized and stored. A CD quality audio requires 16-bit sampling at 44.1 KHz. We may require even higher audiophile rates like 24-bit at 96 KHz. A simple example to think about is the fact that a 1minute of uncompressed mono CD quality audio may require around 5 Mb while a 1 minute of uncompressed stereo CD quality audio will require up to 10 Mb. These are usually compressed in MP3, AAC, Flac, or Ogg Vorbis formats.

Analog video is usually captured by a video camera and then digitized. A variety of analog and digital video formats exist. Raw video can be regarded as being a series of single images with typically 25, 30 or 50 frames per second. As a simple practical example, a 512×512 size monochrome video images take 25×0.25 = 6.25Mb for a minute to store uncompressed video. A typical PAL digital video which is 720×576 pixels per colour frame requires 1.2×25 = 30Mb for a minute to store uncompressed video. A high definition DVD of 1440×1080 = 1.5 Mega-pixels per frame needs 4.5×25 = 112.5Mb for a minute to store uncompressed video. This is not the limit. There are even higher possible frame rates for video processing. Needless to say, for practical use and handling, digital video clearly needs to be compressed otherwise the capacities of the storage units and memory units will be unthinkably impractical.

Chapter Nineteen

Approach to Multimedia Delivery

19.1 Introduction

The Node Gain Scores (NGSs), used as a basis for shaping the max-heap overlay, are arbitrarily determined by the respective bandwidth-latency-products. Constructing a max-heap-form overlay tree governed by the magnitudes of individual Node Gain Scores (NGSs), each earned as a synergy of the discrepancy ratio of the bandwidth requested with respect to the estimated available bandwidth, and the latency discrepancy ratio between the nodes and the source node, plays a great role in reducing a vital induced packet loss caused by, otherwise, the schemes which do not consider these parameters on placing the nodes on the overlay trees.

It is, here in this work, proposed that each node to be positioned according to the NGS it earns from a function governed by the four main influencing parameters—the estimated available bandwidth, B_a; the individual node's requested bandwidth, B_r; the proposed node latency to its prospective parent, L_p; and the suggested best latency as advised by the source node, L_b. The NGS of each node is pre-calculated as an integrated measure from a fraction of the bandwidth discrepancy ratio (BDR) and that of the latency discrepancy ratio (LDR) with the weights of α and $\beta=(C-\alpha)$, respectively, and with arbitrary chosen α ranging between 0 and C, that is $0 \leq \alpha \leq C$, and $\beta = C - \alpha$ to make sure that the NGS values, used as node IDs, maintain a good possibility of uniqueness and a good balance between the BDR and the LDR—whichever is the most critical factor and vice versa.

The constant C is, hence, chosen depending on the expected unique Node_IDs desired. A max-heap-form tree is constructed with an assumption that all the nodes possess, as it must practically be, NGS less than the source node. To maintain a sense of load balance, the children of each level's siblings are evenly distributed such that a node can not accept a second child, and so on, until all its siblings able to do so, have already acquired the same number of children, and that is so logically done from left to right in a conceptual overlay tree.

The records of the pair-wise approximate available bandwidths as measured by a pathChirp scheme at individual nodes are maintained and the evaluation measures as compared to other schemes like Bandwidth Aware overlay multicaSt architecture (BASE), Tree Building Control Protocol (TBCP), and Host Multicast Tree Protocol (HMTP) have been conducted.

When a moderate sized overlay multicasting group is under worst case consideration, this new scheme seems to generally perform better in terms of trade-off between packet delivery ratio which means a reduced packet loss; the maximum link stress; the acceptable control overhead; and the reasonable end-to-end delays.

19.2 Overlay multicast induced packet loss

Conventional overlay multicast trees have been customarily push packet data from top to down. It is a common and simpler practice to design an overlay scheme that the source node transmits media down to its immediate children and these children to recursively relay the media to their immediate children downwards. This goes well if and only if the upper placed nodes on the tree performs better than their lower counterparts. Otherwise, if the upper nodes experience any sort of packet loss, this loss is carried on and induced down the tree. This is what we call the overlay multicast induced packet loss.

One very critical agent of induced packet loss spread has been relying on the idea that the new nodes arriving should tail on a tree at the position of the leaf nodes after abiding with a simple algorithm under context. When this new joined leaf-positioned nodes need a bandwidth and other performance metrics more than what the assigned parent can provide, there every chance that the packet loss will be introduced. This is going to further introduce more loss to the incoming children, grandchildren, and so on. Available bandwidth, latency, suggested latency, and requested bandwidth, if well incorporated, can reduce at large the effect of the overlay multicast induced packet loss. NGS-based scheme relying on the function derived from the mentioned parameters has been shown to be one of the successful mechanism to overcome with the induced packet loss in overlay multicast trees.

19.3 Research approach and methodology

19.3.1 The Problem Statement

Low packet delivery ratio in a heavily loaded traffic has been one of the critical issues requiring intensive scheme while dealing with such possessing systems at, especially, the worst cases. The applicability of most overlay schemes is barely limited to light traffic loads. Dealing with heavy loads worst cases of traffic using the existing schemes is surely prone to a lot of packet loss. The case becomes more alarming when we think of the induced packet loss where the higher leveled nodes in an overlay spread the loss they have incurred to their respective lower leveled nodes (children and grandchildren).

The proposal under this work aims at addressing the need of devoting the available and simple measurable parameters in creating a better overlay multicast tree which can utilize the same available resources optimally to prevent or at least to largely reduce the induced packet loss for the purpose of achieving higher packet delivery ratio at the end nodes.

The main interest being to preserve delivery failure influences by other higher level nodes in an overlay construction. In the worst case, it is not uncommon to experience a lot of frequently cut downs of the service if a number of end users wanting to access the server exceeds the allowable capability of the content server. In this case, it is wiser to spend more time and to introduce comparatively higher overhead in return for a rather stable delivery overlay tree with acceptable higher packet delivery ratio at the end nodes.

19.3.2 Selected Approach to the Problem

It is proposed that each node be positioned according to the NGS it earns from a function governed by the four main influencing parameters—the estimated available bandwidth, B_a; the individual node's requested bandwidth, B_r; the proposed node latency to its prospective parent, L_p; and the suggested best latency as advised by the source node, L_b. On optimizing the NGS function in which the NGSs will be used as the Node_IDs, it has been found that it is a direct proportion of the said parameters as expressed in the relation:—*NGS* α *[(B_a - B_r), 1/B_r, (L_b - L_p), 1/ L_p)]* The NGS of each node is pre-calculated as an integrated measure from a fraction of the bandwidth discrepancy ratio *(BDR)* and that of the latency discrepancy ratio *(LDR)* with the weights of α and *(C - α)*, respectively and with arbitrary chosen α ranging between 0 and *C* to make sure that the NGS values, used as node IDs, maintain a good possibility of uniqueness and a good balance between the most critical factor among the BDR and LDR as described by a constant *C*.

The NGSs are expected to be unique nearest integers such that if two or more nodes possess the same NGS values, the NGS of the newest node is recursively decreased by a numeric one and so on. A max-heap-form tree is then constructed with an assumption that all the nodes possess, as it must practically be, NGSs less than the source node's NGS. This scheme tries to maintain a sense of load balance by evenly distributing the children of each level's siblings such that a node can not accept or can not be assigned the nth child until all its siblings have been able to register (n-1) children if they are capable of doing so as dictated by their out-degree boundaries and their bandwidth capabilities. That assignment is so logically done from left to right in a conceptual overlay tree construction.

The record of the estimated available bandwidths as measured by a pathChirp scheme [8], by Ribeiro et al, at individual nodes is maintained. The two parameters, B_r and L_p, are fed to the source node from individual nodes as per individual nodes requirements for better delivery of the content. The available bandwidth, *Ba,* and the source-based proposed best latency available, L_b, are the measured ones through probing. Comparing to other schemes like BASE, the evaluation measures conducted with a moderate sized overlay multicasting group under the worst case consideration, with trade-off, seem to generally perform better in terms of trade-off between packet delivery ratio, maximum link stress, control overhead, and end-to-end delay. To achieve NGS-based proposal, we form an overlay multicast tree (OMT) as max heap form depending on calculated NGS in such away that a member decides the level and position on an OMT and also the degree by comparing its own NGS and other group members' NGS. This degree is dynamically and periodically adjusted according to the current amount of available bandwidth, with respect to the requested one, and hence it is having a direct link with the current values of the NGS. The NGS value seems to be an extract of the bandwidth-latency products and therefore the scheme to reduce induced packet loss is basically an orientation of max-heap-form overlay construction based on the bandwidth-latency products.

Chapter Twenty

Bandwidth-Latency-Product Measures (BLPMs)

20.1 Introduction

Constructing max-heap-form overlay governed by Node Gain Scores (NGSs) earned from requested bandwidth discrepancy ratio (BDR) with respect to available bandwidth, and latency discrepancy ratio (LDR) between the nodes and the source, reduces induced packet loss caused by schemes neglecting these parameters in constructing the tree. This proposal positions nodes using NGS function governed by available bandwidth, requested bandwidth, proposed latency to prospective parent, and best latency as advised by the source. BDR and LDR are set with α and $(C-\alpha)$ respective weights, and arbitrarily chosen α between 0 and C. NGSs, used as Node_IDs, maintain uniqueness and balance between the critical factors. Max-heap-form tree is constructed assuming that all the nodes possess NGS less than the source. For load balancing, parent-node can not accept another child until all its able siblings acquire same number. This scheme outperforms the others, like BASE, in packet delivery, link stress, and end-to-end delay.

20.1.1 BLPM scheme overview

Low packet delivery ratio in a heavily loaded traffic has been one of the critical issues requiring intensive scheme while dealing with such possessing systems at, especially, the worst cases. The applicability of most overlay schemes is barely limited to light traffic loads. Dealing with heavy loads worst cases of traffic using the existing schemes is surely prone to a lot of packet loss. The case becomes more alarming when we think of the induced packet loss where the higher leveled nodes in an overlay spread the loss they have incurred to their respective lower leveled nodes (children and grandchildren). The proposal under this work aims at addressing the need of devoting the available and simple measurable parameters in creating a better overlay multicast tree which can utilize the same available resources optimally to prevent or at least to largely reduce the induced packet loss for the purpose of achieving higher packet delivery ratio at the end nodes. The main interest being to preserve delivery failure influences by other higher level nodes in an overlay construction. In the worst case, it is not uncommon to experience a lot of frequently cut downs of the service if a number of end users wanting to access the server exceeds the allowable capability of the content server. In this case, it is wiser to spend more time and to introduce comparatively higher overhead in return for a rather stable delivery overlay tree with acceptable higher packet delivery ratio at the end nodes. The magnitudes of individual Node Gain Scores (NGSs) are the basis for constructing a max-heap-form overlay tree which will prevent the spread of packet loss caused by assigning the higher resourced nodes as children of the lower resourced ones. The NGS, a measure of an overall node's strength, is calculated from a synergy of the discrepancy ratios of the bandwidth requested with respect to the estimated available bandwidth to serve that node, and of the latency discrepancy ratios between the nodes and the best latency available with a source node. The main contribution of this proposal is to play a great role in reducing a vital induced packet loss caused by the present schemes which do not consider these parameters on placing the nodes on the overlay tree.

It is proposed that each node be positioned according to the NGS it earns from a function governed by the four main influencing parameters—the estimated available bandwidth, B_a; the individual node's requested bandwidth, B_r; the proposed node latency to its prospective parent, L_p; and the suggested best latency as advised by the source node, L_b. The NGS of each node is pre-calculated as an integrated measure from a fraction of the bandwidth discrepancy ratio *(BDR)* and that of the latency discrepancy ratio *(LDR)* with the weights of α and *(C - α)*, respectively and with arbitrary chosen α ranging between 0 and 1,000 to make sure that the NGS values, used as node IDs, maintain a good possibility of uniqueness and a good balance between the most critical factor between the BDR and the LDR.

The NGSs are expected to be unique nearest integers such that if two or more nodes possess the same NGS values, the NGS of the newest node is recursively decreased by a numeric one and so on. A max-heap-form tree is then constructed with an assumption that all the nodes possess, as it must practically be, NGSs less than the source node's NGS. This scheme tries to maintain a sense of load balance by evenly distributing the children of each level's siblings such that a node can not accept or can not be assigned the nth child until all its siblings have been able to register (n-1) children if they are capable of doing so as dictated by their out-degree boundaries and their bandwidth capabilities. That assignment is so logically done from left to right in a conceptual overlay tree construction. The record of the estimated available bandwidths as measured by a pathChirp scheme [8], by Ribeiro et al, at individual nodes is maintained. The two parameters, B_r and L_p, are fed to the source node from individual nodes as per individual nodes requirements for better delivery of the content. The available bandwidth, Ba, and the source-based proposed best latency available, L_b, are the measured ones through probing. Comparing to other schemes like BASE, the evaluation measures conducted with a moderate sized overlay multicasting group under the worst case consideration, with trade-off, seem to generally perform better in terms of trade-off between packet delivery ratio, maximum link stress, control overhead, and end-to-end delay. To achieve NGS-based proposal, we form an overlay multicast tree (OMT) as max heap form depending on calculated NGS in such away that a member decides the level and position on an OMT and also the degree by comparing its own NGS and other group members' NGS. This degree is dynamically and periodically adjusted according to the current amount of available bandwidth, with respect to the requested one, and hence it is having a direct link with the current values of the NGS. The NGS value seems to be an extract of the bandwidth-latency products and therefore the scheme to reduce induced packet loss is basically an orientation of max-heap-form overlay construction based on the bandwidth-latency products.

20.1.2 Related Works

20.1.2.1 Preamble

Network and node congestion are the causes of higher link stress which leads to a low link utilization and, worse enough, to unwanted nodal packet loss and induced packet loss. In overlay multicasting tree, induced packet loss occurs when one group member (higher leveled) does not receive data packet. Many research works have been in progress to try to prevent packet loss by avoiding the said two types of congestions. The other alternatives have been concentrating on recovering the lost packets. Recently, the focus has been on creating multi-path routes to deliver packet which can not be done through one path or when one path fails. The Probabilistic Resilient Multicast (PRM) proposed by Banerjee et al in [3] guarantees arbitrarily high data delivery ratio and low latency bounds in proactive or reactive manners while in [4], Xie et al have proposed a scheme for a node to receive data not only from its parent, but also from some other nodes in the tree and this seems to reduce induced packet loss. These schemes together with other schemes so far proposed—like Nemo proposed by Birrer et al in [5], ROMA suggested by Kwon et al in [6], and so on—have been largely addressing the light load issues. For heavy traffic loads, the recent schemes have been displaying a great failure. None has been considering the available bandwidth as the key metric to construct and maintain overlay multicast tree which avoids congestion at the network and at the nodal levels. Such schemes that can deal well with the induced packet loss include the schemes that make use of the available bandwidth to construct their overlays.

K. K. To et al in [9] have come up with an extended NICE scheme, P-NICE, which introduces the routing of application-layer multicast packets over multiple paths in the network. To compensate for the increased control overheads, they have used a new algorithm to smooth out the measured RTT and to filter short-term random variations to reduce the number and frequency of topology rearrangements. Even though P-NICE seems to substantially outperform NICE in high-data-rate applications, the scheme did not consider the discrepancy between the latencies demanded and that best available. Neither did the discrepancies in terms of the requested and the available bandwidths have been taken seriously.

20.1.2.2 Available bandwidth based overlays

In [1], K. Kim has introduced a scheme named BASE (Bandwidth Aware overlay multicaSt architecturE) and explained the work as an effort to minimize link stress by connecting each link over overlay multicast tree with available bandwidth metric. In this proposal, Kim has suggested that BASE can minimize group member's delivery failure influence by locating group members with more available bandwidth at higher level on overlay multicast tree. In doing so, Kim has observed that BASE can obtain higher packet delivery ratio because packet delivery failure will rarely happen at higher levels on overlay multicast tree. Kwan et al in [2] have addressed a major challenge in designing application layer multicast protocols by improving the joining and maintenance procedures of an overlay multicast tree. They have addressed the challenge by speeding up the formation of the tree and by enhancing the efficiency of the tree maintenance and they have taken both the bandwidth availability and the round-trip-time (RTT) into consideration when a newcomer selects its parent node, but they did not consider the relationship between the discrepancy of the latencies affordable with respect to the one to be offered. They have neither considered the discrepancy between the requested bandwidth and the availability of the bandwidth. These constraints are of critical importance especially when the overlay under discussion is constituted of mainly the worst case multicasting groups.

In [7], Zhu et al have presented a bandwidth-efficient scheme called CRBR that seamlessly integrates network access control and group key management. This scheme seems to incur much smaller communication overhead than two other well-known schemes when they are directly applied in overlay multicast, but again the efficiency is just based on the bandwidth, without taking into consideration the other critical parameters. Further more, Yang et al in [10] have proposed a scheme with key metrics judging the quality of a multicast tree being the normalized aggregate delay, D, the normalized aggregate cost, B and the weighted sum of delay and bandwidth consumption (WSDB). The individual discrepancies of bandwidths and latencies have not, though, kept into serious consideration. All the above schemes do not, or just consider, the available bandwidth as the core metric in positioning the overlay multicasting nodes in a tree. These can heuristically induce tremendous packet loss as the tree gets deeper with bandwidth demand of the nodes randomly distributed.

The fundamental max-heap overlay schemes have been thought to partially solve the bandwidth demand challenge of individual nodes. In yet another recent research, M. Hosseini et al in [12] have based their work on three dynamic network metrics (available bandwidth, latency, and loss) and the two main components to adaptation process. They have devised a mechanism to detect poor performing current parents and/or the determination for parents' switch on hosts. Rather than considering comparative latencies and comparative bandwidths, they have just considered the absolute values which do not help much in reducing the induced packet loss than the proposal of bandwidth-latency product max-heap form construction described in this research work. The aspect of dynamic bandwidth estimation, to serve as an important basis for performance optimization of real-time distributed multimedia applications, has been stressed by Wang et al in [13]. They have developed a bandwidth estimation algorithm for the fast fluctuated internet and analyzed the relationship between the one way delay and the dispersion of packets train. Their work is based on the proposal of an available bandwidth estimation algorithm using the two features while eliminating administrative access to the intermediate routers along the network path. For robustness and efficiency, the top-down approach has been used to infer available bandwidth, but nothing about the BDR and/or LDR has been taken as a basis for the overlay tree construction.

20.1.2.3 Fundamental max-min-heap or rate overlays

The work by Y. Cui et al in [11] has proposed a need of finding the max-min rate allocation in overlay multicast, which is pareto-optimal in terms of network resource utilization. They have presented a distributed algorithm, which is able to return the max-min rate allocation for any given overlay multicast tree. They have also briefly described the mechanism of finding the optimal tree, whose max-min rate allocation is optimal among all trees and, after proving its NP-hardness, they have proposed a heuristic algorithm of overlay multicast tree construction. Yet in another effort to provide an improved overlay multicasting scheme which can reduce packet loss, Kim in [1] has tried to minimizes link stress by mainly connecting each link over overlay multicast tree in a max-heap form with available bandwidth metric used as the overlay nodes' IDs. By an obvious reason, members with more available bandwidth are allocated at higher levels on the tree. In [20], K. Kim and S. Kim, while constructing overlay DDT, they have introduced a max heap structure consisting only of group members' key value measured by mobility and remaining battery, in the case of ad hoc. In their proposal, they have suggested that the less dynamic network a host is placed in, and the more available battery remains, the higher the value. The idea falls short as it is required to analyze the effect of each parameter and combine them properly. The NGS values proposed in this work, based on bandwidth-latency product, at a go, integrates all the four critical parameters to reduce the induced packet loss in an overlay construction.

20.2 Bandwidth-latency-product formulation and interpretation

Recalling the main integrated measure of the overlay construction based on the max-heap form, the NGS, we write the expression in Eq. (20.1) and Eq. (20.2):

$$\text{NGS} \propto [(B_a - B_r), 1/B_r, (L_b - L_p), 1/L_p)]. \dots\dots\dots\dots (20.1)$$

$$NGS = NGS_1 + NGS_2 \dots\dots\dots\dots (20.2)$$

which upon substitution leads to a complete expression for NGS including the constants and the variables and it is given as in Eq. (20.3) or Eq. (20.4):

$$NGS = \frac{\alpha(B_a - B_r)}{B_r} + \frac{\beta(L_b - L_p)}{L_p} \dots\dots\dots (20.3)$$

$$= \frac{\alpha(B_a - B_r)}{B_r} + \frac{(C - \alpha)(L_b - L_p)}{L_p} \dots\dots\dots (20.4)$$

Upon a minor simplification and re-arrangement, the above NGS function can be expressed as in the equation given by Eq. (20.5):

$$NGS = \frac{\alpha L_p B_a + C L_b B_r - C L_p B_r - \alpha L_b B_r}{L_p B_r} \dots\dots\dots (20.5)$$

Now, the simplified and re-arranged NGS can be visualized as the sums and differences of the products of the bandwidths and the latencies. We can safely say that the NGS is a directly increasing function of $\alpha L_p B_a$ and $L_b B_r$. It is also a directly decreasing function of $L_p B_r$ and $\alpha L_b B_r$. The NGS is also a directly decreasing function of $L_b B_r$. The fact is strongly true as we always and practically deal with only the positive bandwidths and the positive latencies. Therefore, the NGS measure is, at the user point of view, nothing but the bandwidth-latency product parameters. This can be quite easily controllable and the trade-off can be simply undergone by choosing the parameters of individual importance.

To test the applicability of the NGS function, we do simple and obvious mathematical tests which can comfortably be extended to general applicability. Since we are interested in positive numerical numbers to stand as node IDs, the main numerical focus in estimating the NGS function is to make sure that the combinations of the ranges of the variables will never result into a negative NGS values for the practical possible values in use of B_a, B_r, L_p, and L_b. That being into consideration, we start by estimating the first component of the NGS, that is NGS$_1$, at the minimum expected B_r of 20MB and setting the value "1" equal to 1,000. If the weight of importance for each of the BDR and the LDR is equally balanced, that is $\alpha = ("1" - \alpha) = 500$, and if we consider the said B_r, we can express the NGS$_1$ as in the equation Eq. (20.6):

$$NGS1_{B_r=20MB}(B_a) \rightarrow 25B_a - 500 \quad\text{.....................................}\quad (20.6)$$

The above expression implies that by fixing B_r equal to 20MB, the NGS$_1$ component behaves as a linear equation of a variable B_a. The contribution of NGS$_1$ to the whole NGS for the given sample figures can be plotted as in Fig. 20.1, with the expected available bandwidth ranging between 20MB and 100MB.

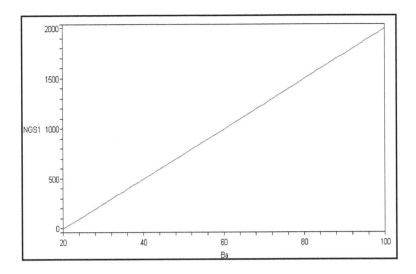

Fig. 20.1. Desired contribution of NGS₁

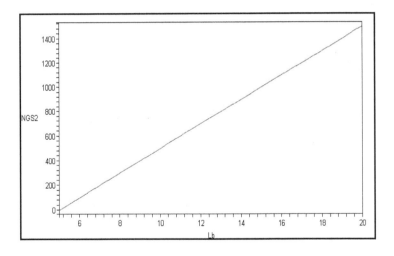

Fig. 20.2. Desired contribution of NGS₂

In a similar fashion, Fig. 20.2 represents the contribution of NGS_2 to the whole NGS. This is obtained by fixing the proposed latency, L_p, to 5ms and allowing L_b to vary between 0 and 20ms which is assumed to be the allowable range for a practical system without any annoyance. We express the second component of NGS as in the formulation shown in Eq. (20.7) with a varying L_b between that range:

$$NGS2_{L_p=5ms}(L_b) \rightarrow 100L_b - 500 \quad\text{------------------------------------}\quad (20.7)$$

Both, the NGS_1 and the NGS_2, show positive monotonic incremental contribution towards the total NGS and therefore, with these sub-functions, we can avoid the possibility of introducing negative node IDs.

On the other hand, we try to contradict the test for the NGS function by considering the undesirable, non-practical values of B_r and L_p, that is, we try to involve the nodes with the requested bandwidth just equal to the possible available bandwidth of 120MB and change the available bandwidth from a minimum of 20MB to that upper limit. The expressions from NGS_1 and NGS_2 which are undesirable are as given below with L_p set at 25ms which is beyond the expected convenient latency of 20ms. The value of L_b is varied between 0 to 20ms which is practically viable and acceptable. We, therefore, have Eq. (20.8) and Eq. (20.9):

$$NGS1_{B_r=120MB}(B_a) \rightarrow \tfrac{25}{6}B_a - 500 \quad\text{--------------------------------}\quad (20.8)$$

$$NGS2_{L_p=25ms}(L_b) \rightarrow 20L_b - 500 \quad\text{------------------------------------}\quad (20.9)$$

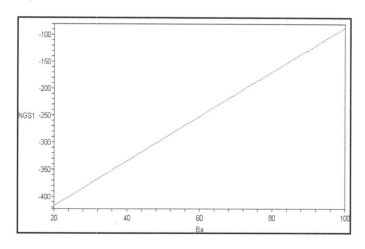

Fig. 20.3. Undesired contribution of NGS_1

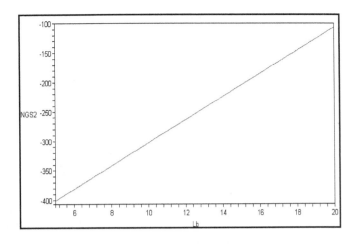

Fig. 20.4. Undesired contribution of NGS₂

Fig. 20.3 and Fig, 20.4, respectively, illustrate the inapplicability of the contributions of NGS₁ and NGS₂ at the contradicted values of test. They both provide negative numerical values which are not desired for the node IDs demarcation. Therefore, the applicability of the proposed NGS function is practically viable under the application ranges of the parameters.

The components-combined NGS expressions for the desired and the undesired ranges of B_r and L_p are, respectively, given in Eq. (20.10) and Eq. (20.11):

$$NGS_{B_r=20MB,L_p=5ms}(B_a,L_b) \rightarrow 25B_a - 1000 + 100L_b \dotfill (20.10)$$

$$NGS_{B_r=120MB,L_p=25ms}(B_a,L_b) \rightarrow \frac{25}{6}B_a - 1000 + 20L_b \dotfill (20.11)$$

The matrices representing the expressions for $NGS_{Br=20,Lp=5}$ and for $NGS_{Br=120,Lp=25}$ are illustrated in Fig. 20.5 and Fig. 20.6, respectively. The former being the positive NGS values which indicates, for practical ranges of B_a and L_b, that we can have up to 3,500 unique whole numbers to represent individual nodes in an overlay multicasting construction. On the contrary, Fig. 20.6 displays all negative, undesired whole numbers for similar ranges of B_a and L_b.

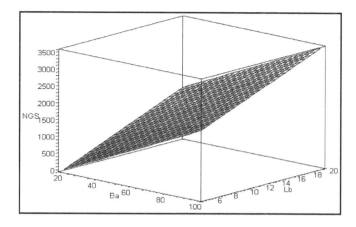

Fig. 20.5. Desired NGS's B_a - L_b matrix

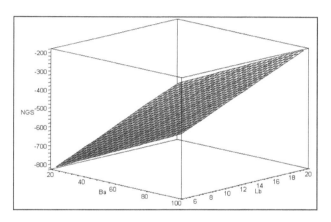

Fig. 20.6. Undesired NGS's B_a - L_b matrix

The above resulting 3-D matrices suggest that we can have a comfortable unique assignment of up to 3,500 nodes in our overlay multicast tree without any ambiguity if we incline our simulation based on the desired NGS's B_a-L_b matrix combination.

Chapter Twenty One

NGS-Based Max-Heap Overlay Multicast Scheme

21.1 Limitations of bandwidth-only-based scheme

In his proposal in [1], Kim has stressed on the available bandwidth as the key metric for the construction and maintenance of an overlay multicast tree, where the higher bandwidth possessing nodes take the upper stake of the overlay tree and those with less bandwidth descend down the overlay tree. The main and sole issue attended by Kim's design has been quick clients' services to reduce or avoid node, network, and path congestions. Taking into consideration the situation with frequent break down of paths and the situation with varied available bandwidth and latencies of the participating nodes, it can be clearly observed that there are other metrics which have not been considered by Kim, in spite of the fact that the integration of them is very important.

The integration of the requested bandwidth from a newly joining member to its expected parent, B_r, the preferred latency from the newcomer to an expected parent, L_p, and the proposal of the best latency available from the newcomer to a proposed parent, L_b, is of equal importance for consideration while constructing and reconstructing an overlay multicast tree with an aim of reducing induced packet loss to the end users. Therefore, the novel proposed scheme in this work aims at building a max-heap form of an overlay tree where the major key of nodes' positioning is a value, NGS, determined by a function of integrated metrics of B_r, L_p, L_b, and B_a.

21.2 The proposed NGS scheme

The bandwidth-latency product oriented max-heap-form overlay construction scheme intends to reduce the incurred induced packet loss and hence improve the packet delivery ratio by imposing integrated metric values, Node Gain Scores (NGSs), acting as node IDs for the nodes positioning, in such away that the higher NGS-ed nodes logically take upper-to-down and alternating left-to-right positions. The reason for the upper-to-down placement is to make sure that the higher metric valued serve the lower ones so that the packet loss causing nodes do not recursively influence the lower children nodes. On the other side, the logical alternating left-to-right approach is for the sake of ensuring load balancing throughout the overlay tree.

21.3 Node gain score (NGS) function

Without loss of generality of the basic aforementioned components, the Node Gain Score (NGS) has, in this paper been defined in terms of two critical components; the bandwidth discrepancy ratio (BDR) and the latency discrepancy ratio (LDR). The α scaled component associated with BDR, NGS_1, is the ratio of the difference between the requested bandwidth from the available bandwidth to the requested value of the bandwidth. NGS_1 is mathematically expressed as in Eq. (21.1):

$$NGS_1 = \frac{\alpha(B_a - B_r)}{B_r} \text{-----------------------------------} (21.1)$$

where α is the constant determining the weight in importance of the BDR factor as compared to the LDR parameters, B_a is the available bandwidth as estimated by the pathChirp mechanism for the path that is suggested to connect that particular node, and B_r implies the requested bandwidth that a node under consideration feels comfortable to be served on. In a similar fashion, NGS_2 is an NGS component describing the weighted importance of the said LDR. With a logical numerical figure '1' standing for any complete set value as per the proportion of the projected number of members, the scale $(1- \alpha)$ is the representation of the factor which an LDR contributes to the computation of the overall NGS. If the proposed latency which a node prefers is L_p and the best latency that the node under context can be offered is L_b, then NGS_2 can be expressed as in the equation Eq. (21.2):

$$NGS_2 = \frac{\beta(L_b - L_p)}{L_p} = \frac{(C - \alpha)(L_b - L_p)}{L_p} \text{-------------------} (21.2)$$

Having described the two main components dictating the positioning of the nodes in a max-heap form overlay tree, we can, hence, simply integrate the two NGS functions into a general NGS function as a sum of the two components and hence write it down as in Eq. (21.3):

$$NGS = NGS_1 + NGS_2 \quad\text{...}\quad (21.3)$$

With NGS_1 and NGS_2 being known, we substitute the respective expressions and obtain the integrated form of NGS as in the following NGS function as in Eq. (21.4) or Eq. (21.5):

$$NGS = \frac{\alpha(B_a - B_r)}{B_r} + \frac{\beta(L_b - L_p)}{L_p} \quad\text{.............................}\quad (21.4)$$

$$= \frac{\alpha(B_a - B_r)}{B_r} + \frac{(C-\alpha)(L_b - L_p)}{L_p} \quad\text{.....................}\quad (21.5)$$

Therefore, the NGS value of each node can be collectively determined by the two main components; the BDR and the LDR, scaled by arbitraries α and $(C - \alpha)$ as their factors and with a consideration of the possibility of generating as much unique number as required of NGS values which will stand as the unique node IDs on the max-heap form overlay tree.

21.4 Logical member positioning in NGSs based overlay

21.4.1 Max-heap form overlay tree construction

Based on the NGS values as calculated by the NGS function, the members are ranked from top to down and alternating from left to right in such away that the overlay tree constructed maintains a max-heap form. Therefore, the max-heap form of the members in overlay tree is mainly determined by the products of the bandwidths and latencies as illustrated by the NGS's general and simplified expression. In this proposal, we introduce restricted-but-not-fixed out-degree allocation of paths. The numbers of out-degrees will mainly depend on the values of NGS the nodes earn, and hence will depend on the BDR and LDR values which, indirectly, determine the amount and the extent in which those particular nodes can sustain induced packet loss.

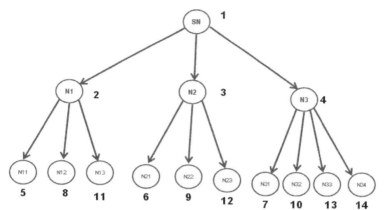

Restricted but NOT fixed out-degrees – depends on the NGS values

$$NGS_{SN} > NGS_{N1} > NGS_{N2} > NGS_{N3} > NGS_{N11} > NGS_{N21} > NGS_{N31} > NGS_{N12} > NGS_{N22} > NGS_{N32} > NGS_{N13} > NGS_{N23} > NGS_{N33} > \ldots$$

Fig. 21.1. Logical member positioning in NGS based overlay—
NGS based members positioning

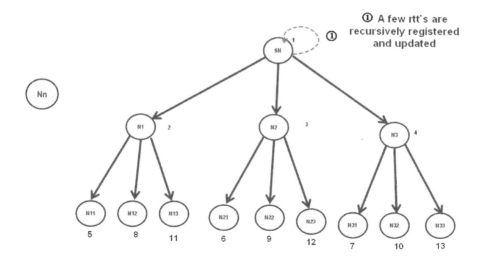

Fig. 21.2. Logical member positioning in NGS based overlay—
Intermediate step before N_n joins

The max-heap form suggested in this design can be compared with the heap form described in Fig 21.1 and the positioning of the members numbered by considering their NGS values they have earned as a result of their BDRs and LDRs. The members are alternated across the level in such away that each sibling at a given level evenly serves equal number of children before the others serve more, unless that particular sibling has no capability to hold more. In Fig. 21.1, a source node, SN, logically takes position *"1"* because it has, or assumed to have, the highest NGS value, NGS_{SN}, while nodes N_1, N_2, and N_3 respectively having NGS values NGS_{N1}, NGS_{N2}, and NGS_{N3}, are respectively positioned at positions "2", "3", and "4". From there then, the newly coming nodes are spread evenly to be served as the children of N_1, N_2, and N_3 with positions "5" as a child of N_1, "6" falling under N_2, and N_3 carrying a new node positioned at "7".

The same fashion is followed until when N_1 and N_2 can no longer hold any more children and hence the subsequent new node, N_{34}, is positioned at "14". Fig. 21.2 shows the intermediate stage which a new node N_n, meets. There is an updated list of a few best rtt's registered at the SN which maintain the best L_b values that will be used to offer the new members wanting to join so that the later can use the values to compare with their L_p values. A source node (SN) starts with an rtt to itself registered as 0 and SN stores rtt's of only a few members with remaining out-degree (s) with a consideration that non-fixed out-degree restriction applies. For making it easier in design, we set very large values for the out-degrees, say a nodal out-degree=30! The membership procedure is detailed illustrated shortly in the coming subsection.

21.4.2 Overlay membership based on NGSs

Figs. 21.3 through 21.10 depict the overlay membership based on NGS values of the member nodes.

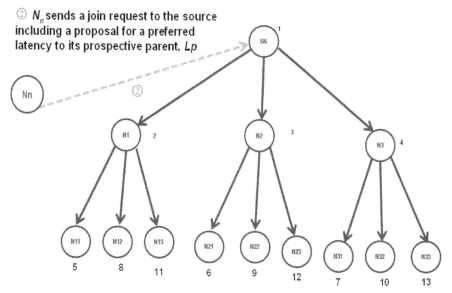

Fig. 21.3. Overlay membership based on NGSs—
a new node, N_n, intends to join a multicast group

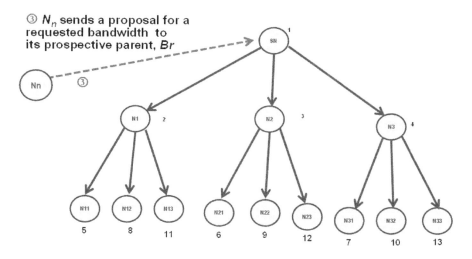

Fig. 21.4. Overlay membership based on NGSs—
N_n sends a proposal for B_r

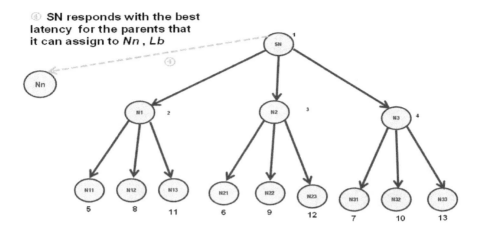

Fig. 21.5. Overlay membership based on NGSs—
SN responds with L_b to N_n.

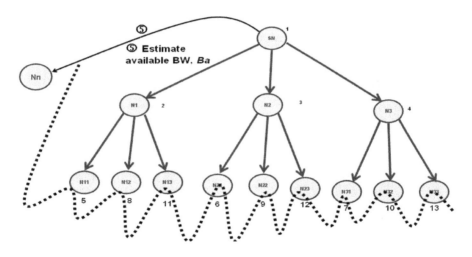

Fig. 21.6. Overlay membership based on NGSs—Applying the
pathChirp mechanism, SN estimates the available bandwidth for the
requesting node, B_a

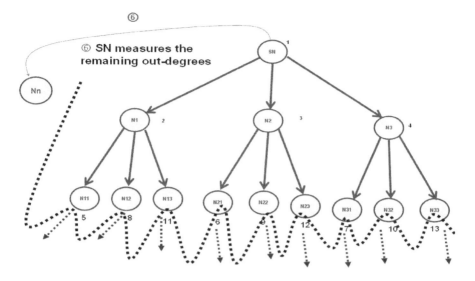

Fig. 21.7. Overlay membership based on NGSs—SN measures the remaining out-degrees and compares with the requesting node's requirements

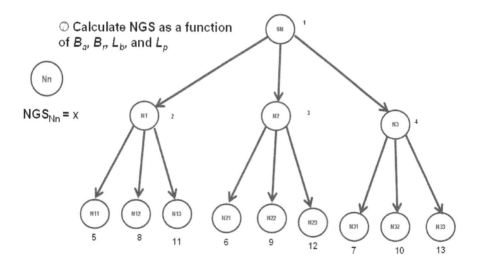

Fig. 21.8. Overlay membership based on NGSs—Node Gain Score (NGS) is calculated as a function of B_a, B_r, L_b, and L_p,

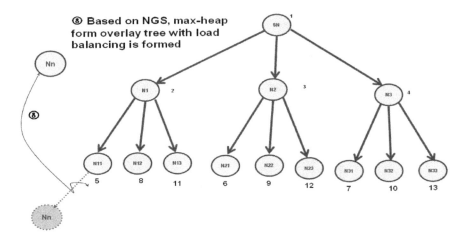

Fig. 21.9. Overlay membership based on NGSs—
a parent node is selected

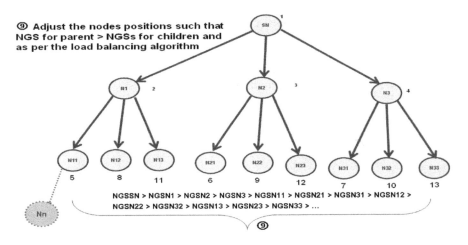

Fig. 21.10. Overlay membership based on NGSs—the adjustment
procedure is applied

In Fig. 21.3, when a new node, N_n, intends to join a multicast group, it sends a join request to the source including a proposal for a preferred latency to its prospective parent, L_p. N_n, as in Fig. 21.4, also sends a proposal for a requested bandwidth to its prospective parent, B_r. Fig.21.5 shows a SN responding with the best latency, L_b, for the parents that it can assign to N_n. This value is the immediate one less than or equal to L_b, if any, else, the smallest value is assigned. In this manner, the algorithm just scans a few rtt's instead of all the existing rtt's. Applying the pathChirp mechanism, in Fig. 21.6, SN also estimates the available bandwidth for the requesting node, B_a, and in Fiq. 21.7, the SN measures the remaining out-degrees and compares with the requesting node's requirements. The Node Gain Score (NGS), in Fig. 21.8, is calculated as a function of B_a, B_r, L_b, and L_p, such that NGS α $[(B_a - B_r),$ $1/B_r,$ $(L_b - L_p),$ $1/ L_p)]$ and that the nearest unique integers are allocated to each node in a tree of overlay multicasting group. As shown in Fig. 21.9, based on the NGS values, a parent node is selected according to the algorithm and based on the maximum NGS heap form of tree with load balancing put into account. Finally, in Fig. 21.10, the adjustment procedure is applied such that the NGS value for a parent is greater than the individual NGS values for the children and the routine to adjust the nodes positions is applied to comply with the inequality $NGS_{SN} > NGS_{N1} > NGS_{N2} > NGS_{N3} > NGS_{N11} > NGS_{N21} > NGS_{N31} > NGS_{N12} > NGS_{N22} > NGS_{N32} > NGS_{N13} > NGS_{N23} > NGS_{N33} > \ldots$ and so on.

Talking about the NGS-based overlay multicast tree maintenance, it is sad, but performance wise paying, that the tree based on bandwidth-latency products should be rebuilt whenever a group member's end-to-end delay exceeds the delay bound as set with a threshold. In addition, we also need to reconstruct the max-heap of the overlay if the available NGS value, based on the BDR and LDR, is less than the threshold set. This is mainly because if the NGS value, and hence the BDR and LDR decrease and the current number of children is not changed, then we can surely experience undesired node congestion. Likewise, if end-to-end delay set is exceeded, a member reinitiates the group join procedure with emphasis that delay should be checked to be within the agreed delay range.

Chapter Twenty Two

Performance Evaluation of Max-Heap Overlay Tree

22.1 Evaluation setup

Transit-stub graph model topologies (GT-ITM) were considered for the ns-2 tool and also the network modeling using OPNET Modeler was considered to create topologies of 1,000 nodes and 128 group members.

The assignment of random link delays of 5-20ms and link capacities of between 20 and 100MB was done. During the process, the available bandwidths were measured by pathChirp scheme. The evaluation measures as compared to BASE mechanism were chosen as packet delivery ratio (pdr) against link capacity traffic (lct), maximum link stress (mls), minimum control overhead (mco), and end-to-end delay (eed).

22.2 Evaluation results and inference

There is, of course, a trade off between adopting this newly designed scheme and sticking to the already available traditional schemes. There are roughly three loss-gain trade-off pairs with this new system: -

Loss 1: A node may experience a lot of pre-join routine check-ups and measurements before being assigned a parent.
Gain 1: Stability is expected to prevail after a node is being accepted to join.
Loss 2: A node may choose a parent with rtt greater than a minimum available, as it just scan for a node with rtt less than or equal to its proposal.
Gain 2: A node saves the time to scan all the rtt's which, sometimes, gives the same results.
Loss 3: A node may join at a parent that is not the best for it, to maintain max-heap NGS form of overlay tree.

Gain 3: A node saves from the possibility of having un-balanced tree, as the mechanism evenly spread the children to the siblings of a given level of the overlay tree.

The comparison simulations of the NGS based mechanism and the BASE scheme gave the results as described in, respectively, Table 22.1 to Table 22.4 and Fig. 22.1 to Fig. 22.4.

Table 22.1. Ratio of percentage packet delivery (%) vs. link capacity traffic (%)

% traffic in link capacity	BASE	NGS
5	97	98.5
10	96.5	98.4
15	95	98
20	94.3	98
25	94.4	97.5
30	94	97.5
35	94	97
40	93.5	97
45	93.2	96.8

% traffic in link capacity	BASE	NGS
50	93.2	96.8
55	93	96.6
60	92.8	96.2
65	92.5	96
70	92.4	96
75	92.3	96
80	92	95.5
85	92	95.4
90	90.8	95.3

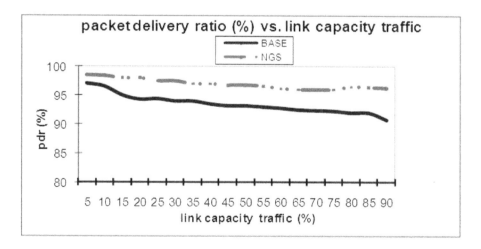

Fig. 22.1. Packet delivery ratio vs. link capacity traffic

Fig, 22.1 shows the comparison results of BASE system and the NGS based overlay constructed for the packet delivery ratios. It can be clearly seen that the proposed NGS mechanism leads to more pdr than its counterpart. The discrepancy in favor of the NGS based bandwidth-latency product max-heap overlay, which increases for the best as the number of nodes increases, seems to have reduced the packet loss tendency which could have otherwise been a must when a simple BASE would have been used.

Table 22.2. Maximum link stress vs. number of nodes

Number of Nodes	BASE	NGS		Number of Nodes	BASE	NGS
5	1.8	1.8		70	3.6	2.4
10	1.8	1.7		75	3.7	2.4
15	1.9	1.7		80	3.7	2.5
20	2	1.7		85	3.7	2.5
25	2	1.9		90	3.9	2.5
30	2	1.9		95	4	2.5
35	2.6	2		100	4	2.5
40	2.7	2.1		105	4.3	2.5
45	3	2.2		110	4.6	2.6
50	3	2.2		115	4.8	2.6
55	3.5	2.3		120	5	2.6
60	3.5	2.4		125	5.2	2.6
65	3.6	2.4		128	5.6	2.6

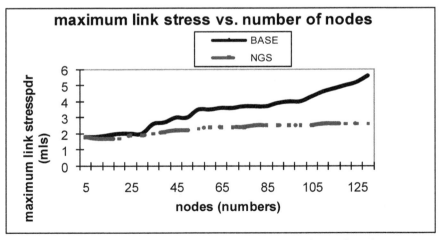

Fig. 22.2. Maximum link stress vs. number of nodes

Similarly, in terms of maximum link stress, the NGS based mechanism seems to do better with the lowest values, as depicted in Fig. 22.2. For the case of the NGS scheme, the maximum link stress remains with a sparingly increasing change when the group size gets bigger. In contrast, the BASE scheme does not tolerate bigger groups in the sense of maximum link stress. It is, again, clear that the NGS mechanism fits better than the BASE, especially for a growing group of overlay multicasting system.

Table 22.3. Minimum control overhead vs. number of nodes

Number of Nodes	BASE	NGS
5	0.5	1.8
10	0.5	1.7
15	0.55	1.7
20	0.55	1.7
25	0.6	1.9
30	0.6	1.9
35	0.6	2
40	0.7	2.1
45	0.8	2.2
50	0.9	2.2
55	0.95	2.3
60	1	2.4
65	1.1	2.4

Number of Nodes	BASE	NGS
70	1.1	2.4
75	1.2	2.4
80	1.3	2.45
85	1.35	2.45
90	1.45	2.45
95	1.45	2.46
100	1.5	2.47
105	1.5	2.48
110	1.6	2.49
115	1.6	2.49
120	1.8	2.7
125	1.8	2.7
128	2	2.7

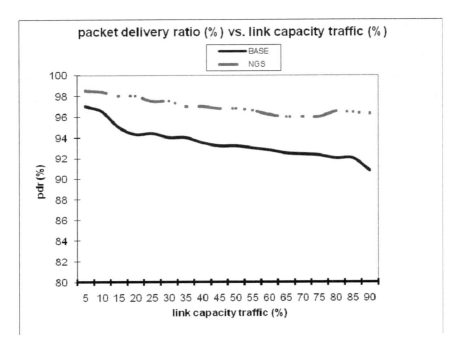

Fig. 22.3. Minimum control overhead vs. Link capacity traffic

In terms of minimum control overhead needed to maintain service, Fig. 22.3 shows that NGS based max-heap overlay does not perform better than BASE, but the bigger the group, the closer the NGS values to the BASE values. In addition, the trade-off between the pdr and the control overhead pays enough as far as the whole system performance is concerned.

Table 22.4. End-to-end delay vs. number of nodes

Number of Nodes	BASE	NGS
5	89.5	70.5
10	82	77
15	125	105
20	110.5	98.5
25	71.5	61.5
30	78.5	72.5
35	102	112
40	159	153
45	131	139
50	151.5	141.5
55	115.6	111.6
60	140	132
65	167.5	162.5
70	178	167

Number of Nodes	BASE	NGS
75	131	122
80	139.5	129.5
85	187.5	182.5
90	110.5	112.5
95	177.5	167.5
100	188	178
105	121	117
110	189.5	180.5
115	187.5	181.5
120	190.5	180.5
125	221	201
126	199.4	179.4
127	219	209.5
128	217.5	207

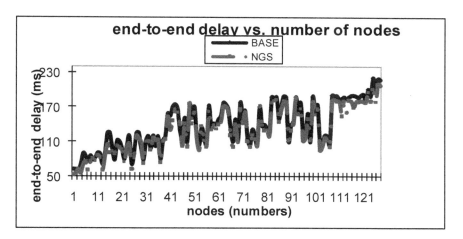

Fig. 22.4. End-to-end delay vs. Number of nodes

Fig. 22.4 indicates that the end-to-end delays encountered by both the systems are almost overlapping with NGS based scheme doing quite better at low values especially when the number of nodes becomes higher.

Respectively, Table 22.5 through Table 22.8 and Fig. 22.5 through Fig. 22.8 provide clear comparisons among BASE, NGS, HMTP, and TBCP mechanisms. There is, of course, a trade off between adopting this newly designed scheme and sticking to the already available traditional schemes. The comparison simulations of the NGS based mechanism with the schemes like BASE, HMTP, and TBCP gave the results as described in Fig. 22.5 through Fig. 22.8. Fig. 22.5 shows the comparison results of BASE, HMTP, and TBCP systems and the NGS based overlay constructed for the packet loss ratios. It can be clearly seen that the proposed mechanism leads to less packet than its counterparts.

Table 22.5. Ratio of percentage packet loss (%) vs. link capacity traffic (%)—comparison of BASE, NGS, HMTP, and TBCP

% traffic in link capacity	BASE	NGS	HMTP	TBCP
5	3	1.5	2.8	2.4
10	3.5	1.6	3.7	4
15	5	2	5.3	4.8
20	5.7	2	6	5.8
25	5.6	2.5	6.2	5.9
30	6	2.5	6.8	6.3
35	6	3	7	6.5
40	6.5	3	7.8	6.8
45	6.8	3.2	7.9	7
50	6.8	3.2	8	7.1
55	7	3.4	8.5	7.3
60	7.2	3.8	8.8	7.6
65	7.5	4	9.1	7.9
70	7.6	4	9.2	8.3
75	7.7	4	9.5	8.5
80	8	4.5	9.7	9
85	8	4.5	9.9	9.2
90	9.2	4.7	9.9	9.4

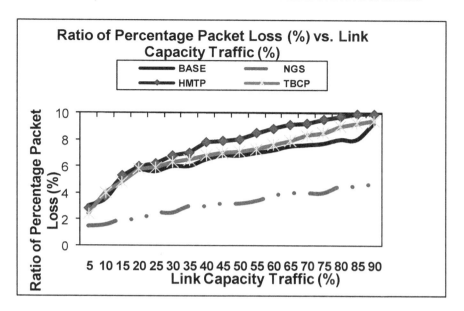

Fig. 22.5. Ratio of percentage packet loss (%) vs. link capacity traffic (%)—comparison of BASE, NGS, HMTP, and TBCP

Table 22.6. Maximum link stress vs. number of nodes—comparison of BASE, NGS, HMTP, and TBCP

Number of Nodes	BASE	NGS	HMTP	TBCP
5	1.8	1.8	2	1.6
10	1.8	1.7	2	1.8
15	1.9	1.7	2.1	2
20	2	1.7	2.2	2
25	2	1.9	2.2	2.3
30	2	1.9	2.3	2.4
35	2.6	2	3	2.6
40	2.7	2.1	3.2	2.5
45	3	2.2	3.3	2.7
50	3	2.2	3.4	2.7
55	3.5	2.3	3.7	3
60	3.5	2.4	3.8	3

65	3.6	2.4	3.8	3.1
70	3.6	2.4	3.9	3.3
75	3.7	2.4	4	3.3
80	3.7	2.5	4	3.3
85	3.7	2.5	4.3	3.4
90	3.9	2.5	4.7	3.5
95	4	2.5	5	3.6
100	4	2.5	5	3.7
105	4.3	2.5	5.4	3.7
110	4.6	2.6	5.6	3.8
115	4.8	2.6	5.8	3.9
120	5	2.6	5.8	3.9
125	5.2	2.6	5.9	4
128	5.6	2.6	5.9	4.1

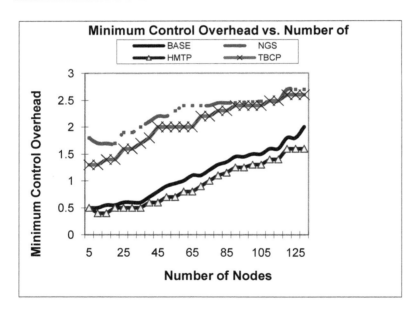

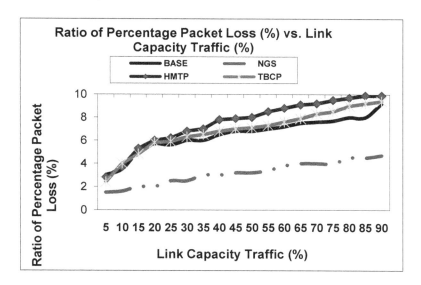

Fig. 22.6. Maximum link stress vs. number of nodes—comparison of BASE, NGS, HMTP, and TBCP

Table 22.7. Minimum control overhead vs. number of nodes—comparison of BASE, NGS, HMTP, and TBCP

Number of Nodes	BASE	NGS	HMTP	TBCP
5	0.5	1.8	0.5	1.3
10	0.5	1.7	0.4	1.3
15	0.55	1.7	0.4	1.4
20	0.55	1.7	0.5	1.4
25	0.6	1.9	0.5	1.6
30	0.6	1.9	0.5	1.6
35	0.6	2	0.5	1.7
40	0.7	2.1	0.6	1.8
45	0.8	2.2	0.6	2

50	0.9	2.2	0.7	2
55	0.95	2.3	0.7	2
60	1	2.4	0.8	2
65	1.1	2.4	0.8	2
70	1.1	2.4	0.9	2.2
75	1.2	2.4	1	2.2
80	1.3	2.45	1.1	2.3
85	1.35	2.45	1.15	2.3
90	1.45	2.45	1.25	2.4
95	1.45	2.46	1.25	2.4
100	1.5	2.47	1.3	2.4
105	1.5	2.48	1.3	2.4
110	1.6	2.49	1.4	2.49
115	1.6	2.49	1.4	2.49
120	1.8	2.7	1.6	2.6
125	1.8	2.7	1.6	2.6
128	2	2.7	1.6	2.6

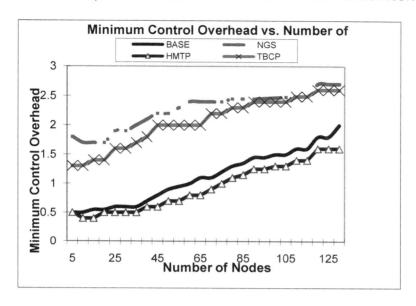

Fig. 22.7. Minimum control overhead vs. number of nodes—
comparison of BASE, NGS, HMTP, and TBCP

Table 22.8. End-to-end delay vs. number of nodes—comparison of
BASE, NGS, HMTP, and TBCP

Number of Nodes	BASE	NGS	HMTP	TBCP
4	70.1	55.5	90.1	75.5
8	85.5	75.5	120.5	105.5
12	88.5	82.5	122.2	112.5
16	115.6	90.5	142.5	131.1
20	110.5	98.5	137.7	126.6
24	120	106	157	128.8
28	125.5	115.5	159.1	143.6
32	119.5	111.5	183.4	141.5
36	119	113	174.5	157.6
40	159	145.1	187.8	173.1

44	160.5	140	200.5	176.7
48	109	100	192.2	182.6
52	128.1	104.5	204	184.8
56	127.4	100.5	200	195.2
60	140	132	213.6	193.7
64	174.9	154.6	219.5	193.7
68	149	139	215.8	190.7
72	163.5	153.2	229.1	195.9
76	143.6	126.6	226.8	202.5
80	139.5	129.5	223.9	198.8
84	184.9	164.9	234.9	194.4
88	189	159.1	239.4	201.1
92	160.5	140	233.5	200
96	170	131.8	243.1	204.7
100	188	156.8	244.5	206.9
104	150	124.4	242.8	205.5
108	190.5	170.8	237.2	210.6
112	185	171.6	245.5	215.1
116	190	176	246.4	217.3
120	190.5	171.6	248.9	222.4
124	198.8	179.5	251.9	228.3
128	217.5	201.8	252	232

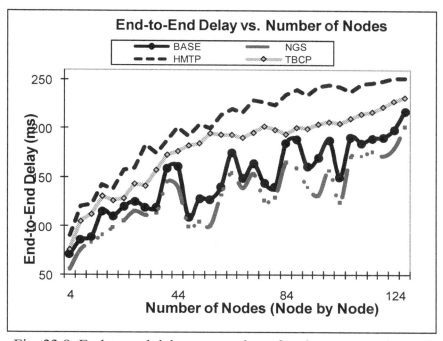

Fig. 22.8. End-to-end delay vs. number of nodes—comparison of BASE, NGS, HMTP, and TBCP

The discrepancy, in favor of the NGS based bandwidth-latency product max-heap overlay, increases for the best as the number of nodes increases. It seems to have reduced the packet loss tendency which could have otherwise been a must when a simple BASE or other traditional mechanisms would have been used. Similarly, in terms of maximum link stress, the NGS based mechanism seems to do better with the lowest values, as depicted in Fig. 22.6. For the case of the NGS scheme, the maximum link stress remains with a sparingly increasing change when the group size gets bigger.

In contrast, the BASE scheme does not tolerate bigger groups in the sense of maximum link stress. It is, again, clear that the NGS mechanism fits better than the BASE and other mechanisms, especially for a growing group of overlay multicasting.

In terms of minimum control overhead needed to maintain service, Fig. 22.7 shows that NGS based max-heap overlay does not perform better than BASE and other mechanisms, but the bigger the group, the closer the NGS's values to the BASE's and other mechanisms'.

In addition, the trade-off between the packet loss and the control overhead pays enough as far as the whole system performance is concerned in enhancing media delivery. Fig. 22.8 indicates that the end-to-end delays encountered by BASE and NGS systems are almost overlapping with NGS based scheme doing quite better at lower values especially when the number of nodes becomes higher. The values are somehow unpredictable due to the nature of the NGS calculations based on the available bandwidths and other parameters. For the topology chosen, HMTP seems to have the worst performance in terms of end to end delays while TBCP comes second.

Chapter Twenty Three

Conclusions, Discussion and Possible Future Direction

23.1 Conclusion

Instead of just considering a general delay metric, the newly proposed NGS, bandwidth-latency products based architecture adapts the available bandwidth metric, the delay metric, the relative rtt metric with respect to a suggested/preferred/proposed rtt, the best delay metric as suggested by the source root to individual members. This makes NGS to enhance the performance by evenly loading the tree and recursively distributing the members with closer NGS on each branch of the network and, hence, reducing the possibility of path congestion or group members congestion by making each group member independently check the congestion situation, considers its position with respect to the source root and other members, and then dynamically adjusts its own variables according to varied traffic environments and its position to quickly reach the source node in case of failure.

23.2 Discussion

The NGS awareness, via the available bandwidth, the requested bandwidth, the proposed latency from the member, and the assigned parent's best available latency have been demonstrated as a very important metric in establishing an overlay which largely reduces the induced packet loss, end-to-end delays, and maximum link stress. It is, however, unavoidable with this scheme to reduce control overhead comparing to other schemes and, therefore, it is needed to device an incorporated mechanism that will maintain the achieved performance with reduced control overhead units.

The limitations of the NGS proposal introduce gains also. Firstly, a node may experience a lot of pre-join routine check-ups and measurements before being assigned a parent for the hope that the stability after being accepted to join will be reliable. Secondly, a node may choose a parent with RTT greater than a minimum available as it just scan for a node with RTT < its proposal. However, a node saves time to scan all the rtt's which, sometimes, give the same results as above or may be less performance. Finally, a node may join at a parent that is not the best for it to maintain max-heap NGS form of tree, but a node saves from the possibility of having un-balanced tree.

23.3 Possible future direction

Since the available bandwidth estimation time may affect the immediate member join, there is a need to device a mechanism such that the estimation is done quicker and adaptively. There should be a mechanism to keep a history of the available paths such that there should be no need to re-estimate the available bandwidth. The minimum control overhead is still higher than the existing BASE system. It is proposed to find a mechanism to reduce that overhead while maintaining other better performances.

References

[1] K. Kim.: Bandwidth Dependent Overlay Multicast Scheme: In International Conference on Communication Systems, 2006. ICCS 2006. 10th IEEE Singapore, IEEE 2006.

[2] T. M. T. Kwan et al.: On Overlay Multicast Tree Construction and Maintenance: In International Conference on Collaborative Computing: Networking, Applications and Worksharing, 2005, IEEE 2005.

[3] S. Banerjee et al.: Scalable Resilient Media Streaming: In NOSSDAV'04, June 16-18, 2004, Cork, Ireland, ACM 2004.

[4] L. Xie et al., An Approach to Reliability Enhancement of Overlay Multicast Trees: In Proceedings of PostGraduate Networking Conference, 2003 PGNet, June 2003.

[5] S. Birrer et al: Resilient Peer-to-Peer Multicast without the Cost. In Proc. of the 12th Annual Multimedia Computing and Networking Conference (MMCN'05), January 2005 (Also published as Tech. Report NWU-CS-04-36).

[6] G. Kwon et al: ROMA: Reliable Overlay Multicast with Loosely Coupled TCP Connections: In Technical Report BU-CS-TR-2003-015, Boston University, 2003.

[7] S. Zhu et al: Efficient Security Mechanisms for Overlay Multicast Based Content Delivery: In Computer Communications 30 (2007), Elsevier B.V., pp 793–806.

[8] V. J. Ribeiro et al., "pathChirp: Efficient Available Bandwidth Estimation for Network Paths," In Proceedings of PAM (Passive and Active Measurement Workshop), Apr. 2003.

[9] K. K. To et al: Parallel Overlays for High Data-rate Multicast Data Transfer: In Computer Networks 51 (2007), Elsevier B.V., pp 31-42.

[10] Y. Jiang et al: A Hierarchical Overlay Multicast Network: In 2004 IEEE International Conference on Multimedia and Expo (ICME), IEEE 2004.

[11] Y. Cui et al: Max-Min Overlay Multicast: Rate Allocation and Tree Construction: In Tech. Report UIUCDCS-R2003-2373/UILU-ENG-2003-1760, 2003.

[12] M. Hosseini et al: End System Multicast Routing for Multi-party Videoconferencing Applications: In Computer Communications 29 (2006), 2005 Elsevier B.V., pp. 2046–2065.

[13] S. S. Wang et al: Fast End-to-End Available Bandwidth Estimation for Real-Time Multimedia Networking: In 8th Workshop on Multimedia Signal Processing, IEEE 2006.

[14] A. Eswaradass et al: Network Bandwidth Predictor (NBP): A System for Online Network performance Forecasting: In Proceedings of the Sixth IEEE International Symposium on Cluster Computing and the Grid (CCGRID'06), IEEE 2006.

[15] A. Habib et al: Incentive Mechanism for Peer-to-Peer Media Streaming: Twelfth IEEE International Workshop on Quality of Service, 2004. IWQOS 2004.

[16] J. A. Strauss: Choosing Internet Paths with High Bulk Transfer Capacity, Department of Electrical Engineering and Computer Science, Massachusetts Institute of Technology (2001), September 2002.

[17] M. Jain et al: End-to-End Available Bandwidth: Measurement Methodology, Dynamics, and Relation with TCP Throughput: In SIGCOMM 2002.

[18] K. Fall et al: The *ns* Manual (formerly *ns* Notes and Documentation): The VINT Project: A Collaboration between researchers at UC Berkeley, LBL, USC/ISI, and Xerox PARC, June 22, 2007.

[19] A. Kim: OPNET Tutorial: OPNET Modeler (OPNET Technologies), March 7, 2003.

[20] K. Kim et al: A Novel Overlay Multicast Protocol in Mobile *Ad Hoc* Networks: Design and Evaluation: IEEE Trans. on Vehicular Tech., Vol. 54, No. 6, Nov. 2005, pp. 2094–2101.

[21] D. Rubenstein et al: Improving reliable multicast using active parity encoding services. In: K. Park (ed.): Computer Networks 44 (2004), Elsevier B.V., pp. 63–78.

[22] C. Y. Lee et al: Reliable overlay multicast trees for private Internet broadcasting with multiple sessions: In Computers & Operations Research 34 (2007), Elsevier Ltd., pp 2849-2864.

[23] A. Busson et al: A new service overlays dimensioning approach based on stochastic geometry: In Performance Evaluation 64 (2007), Elsevier B.V., pp 76–92.

[24] S.Y. Tseng et al: Genetic algorithm for delay- and degree-constrained multimedia broadcasting on overlay networks: In Computer Communications 29 (2006), Elsevier B.V., pp 3625–3632.

[25] R. Cohen et al: The "Global-ISP" paradigm: In Computer Networks 51 (2006), Elsevier B.V., pp 1908-1921.

[26] A. Sehgal et al: A flexible concast-based grouping service: In Computer Networks 50 (2006), Elsevier B. V., pp. 2532–2547.

[27] L. Xiao et al: Mutual anonymous overlay multicast: In Journal of Parallel Distributed Computing 66 (2006), Elsevier Inc., pp. 1205-1216.

[28] T. Braun et al: Explicit routing in multicast overlay networks: In Computer Communications 29 (2006), 2006 Elsevier B.V., pp. 2201–2216.

[29] S. W. Tan et al: A performance comparison of self-organizing application layer multicast overlay construction techniques: In Computer Communications 29 (2006), 2006 Elsevier B. V., pp. 2322–2347.

[30] J. Lee et al: Overlay subgroup communication in large-scale multicast applications: In Computer Communications 29 (2006), 2005 Elsevier B. V., pp. 1201–1212.

[31] L. G. Erice et al: MULTIC: A robust and topology-aware peer-to-peer multicast service: In Computer Communications 29 (2006), 2005 Elsevier B.V., pp. 900–910.

[32] K. Lakshminarayanan et al: End-host controlled multicast routing: In Computer Networks 50 (2006), 2005 Elsevier B.V., pp. 807–825.

[33] H. K. Cho et al: Multicast tree rearrangement to recover node failures in overlay multicast networks: In Computers & Operations Research 33 (2006), 2004 Elsevier B. V., pp. 581–594.

[34] Min-You Wu et al: Placement of proxy-based multicast overlays: In Computer Networks 48(2005), 2005 Elsevier B.V., pp. 627–655.

[35] Minseok Kwon et al: Path-aware overlay multicast: In Computer Networks 47 (2005), 2004 Elsevier B.V., pp. 23–45.

[36] C.K. Yeo et al: A survey of application level multicast techniques: In Computer Communications 27 (2004), 2004 Elsevier B.V., pp. 1547–1568.

[37] N. Wang et al: An overlay framework for provisioning differentiated services in Source Specific Multicast: In Computer Networks 44 (2004), 2003 Elsevier B.V., pp. 481–497.

[38] Gill Waters et al: Optimising multicast structures for grid computing: In Computer Communications 27 (2004), 2004 Elsevier B.V., pp. 1389–1400.

[39] A. Dhamdhere et al: Buffer Sizing for Congested Internet Links: In Proceedings of IEEE INFOCOM 2005, 24th Annual Joint Conference of the IEEE Computer and Communications Societies, IEEE 2005.

[40] S. Fahmy et al: Characterizing Overlay Multicast Networks: In Proceedings of the 11th IEEE International Conference on Network Protocols (ICNP'03), IEEE 2003.

[41] S. J. Wu et al: Improving the Performance of Overlay Multicast with Dynamic Adaptation: In First IEEE Consumer Communications and Networking Conference, 2004. CCNC 2004.

[42] K. Avrachenkov et al: Optimal choice of the buffer size in the Internet routers: In Proceedings of the 44th IEEE Conference on Decision and Control, and the European Control Conference 2005, Seville, Spain, December 12-15, 2005.

[43] S.Y. Shi et al: Routing in Overlay Multicast Networks: In IEEE INFOCOM 2002, IEEE 2002.

[44] Y.H Chu et al: Early Experience with an Internet Broadcast System Based on Overlay Multicast: In USENIX 2004 Annual Technical Conference, General Track, pp. 155-170.

[45] D. Mili et al: Video Broadcasting using Overlay Multicast: In Proceedings of the Seventh IEEE International Symposium on Multimedia (ISM'05), IEEE 2005.

[46] H. R. Shao et al: Adaptive Pre-stored Video Streaming with End System Multicast: TR2002-44, December 2003: In 2002 International conference on Signal Processing.

[47] M. Wang et al: A High-Throughput Overlay Multicast Infra-structure with Network Coding: In: IWQoS 2005, Lecture Notes in Computer Science, Vol. 3552, pp. 37-53, 2005.

[48] J. Zhao et al: LION: Layered Overlay Multicast with Network Coding: In IEEE Transactions on Multimedia, Vol. 8, No. 5, October 2006, IEEE 2006, pp 1021-1032.

[49] W. Tu et al: Worst-Case Delay Control in Multi-Group Overlay Networks: In IEEE Transactions on Parallel and Distributed Systems, IEEE 2007.

[50] R. L. Cruz et al: A Calculus for Network Delay, Part 1: Network Elements in Isolation: In Proceedings of IEEE INFOCOM '88, New Orleans, LA, March 1991.

[51] R. L. Cruz et al: A Calculus for Network Delay, Part 11: Network Analysis: In Proceedings of IEEE INFOCOM '88, New Orleans, LA, March 1988.

[52] A. C. Drummond et al: A Comparison of Measurement-Based Equivalent Bandwidth Estimators: In Globecom 2004 Workshops, IEEE Communications Society, IEEE 2004, pp 320-326.

[53] C. Li et al: A Network Calculus with Effective Bandwidth: In Technical Report, University of Virginia, CS-2003-20, November 2003.

[54] O. Dolejs et al: Optimality of the Tree Building Control Protocol: In Proceedings of the International Conference on Parallel and Distributed Processing Techniques and Applications, CSREA Press, 2002.

[55] L. Mathy et al: An Overlay Tree Building Control Protocol: In J. Crowcroft and M. Hofmann (Eds.): NGC 2001, LNCS 2233, pp. 76-87, 2001.

[56] A. El-Sayed et al: Improving the Scalability of an Application-Level Group Communication Protocol: In 10th International Conference on Telecommunications (ICT'03), Papeete, French Polynesia, February 2003.

[57] R. S. Prasad: Bandwidth estimation: metrics, measurement techniques, and tools: In IEEE Network, 17(6):27--35, November 2003.

[58] Melander et al: A New End-to-End Probing and Analysis Method for Estimating Bandwidth Bottlenecks: In Proceedings of IEEE Globecom, Nov. 2000.

[59] M. Jain et al: End-to-end Available Bandwidth: measurement methodology, dynamics, and relation with TCP throughput: IEEE/ACM Transactions on Networking, Vol. 11, No. 4, August 2003.

[60] J. Strauss et al: A Measurement Study of Available Bandwidth Estimation Tools: In IMC'03, October 27–29, 2003, Miami Beach, Florida, USA, ACM 2003.

[61] S. W. Tan et al: Building Low Delay Application Layer Multicast Trees: In Madjid Merabti and Rubem Pereira, editors, Proceeding of 4th Annual PostGraduate Symposium: The Convergence of Telecommunications, Networking & Broadcasting, pages 27-32. EPSRC, Liverpool John Moore University, June 2003.

[62] P. J. Wagh: DiffServ Overlay Multicast for Videoconferencing: Graduate School: University of Missouri, Columbia, USA.

[63] Y. Cui: Optimal Resource Allocation in Overlay Multicast: In IEEE Transactions on Parallel and Distributed Systems, Vol. 17, No. 8, August 2006, IEEE Computer Society.

[64] J. H Jeon et al: Real-time Adaptive Tree Building in Host-based Multicast Scheme: In: Proceedings of Networking, Sensing and Control, 2005 IEEE.

[65] D.A. Helder et al: End-host multicast communication using switch-trees protocols: In Proceedings of the 2nd IEEE/ACM International Symposium on Cluster Computing and the Grid (CCGRID.02), IEEE2002.

[66] D. Kostic et al: Bullet: High Bandwidth Data Dissemination Using an Overlay Mesh: In *SOSP '03,* October 19–22, 2003, Bolton Landing, New York, USA, 2003 ACM.

[67] D. M. Moen: Overlay multicast for real-time distributed simulation: Defense modeling and simulation Office, Contract NBCH00-02-D-0037, Task Order 0217-001, May 12, 2005.

[68] Y. Cui: Content Distribution in Overlay Multicast: University of Illinois at Urbana-Champaign, Urbana, Illinois, 2005.

[69] H. Hamed: Overlay Multicast Protocols: In 2003 IEEE International Conference on Communications, Vol. 2, Anchorage Alaska, USA.

[70] R. Bhagwan et al: Cone: A Distributed Heap Approach to Resource Selection: In UCSD Tech Report CS2004-0782, University of California, San Diego, USA.

[71] S. B. Moon: Measurement And Analysis Of End-to-End Delay And Loss In The Internet: Department of Computer Science, Graduate School of the University of Massachusetts Amherst, USA, February 2000.

[72] C. Y. Lee et al: An Overlay Multicast to Minimize End-to-end Delay in IP Networks: Dept. of Industrial Engineering, KAIST, Taejon, Korea: In International Conference on Communication Technology, 2006 (ICCT'06), Nov. 2006, pp. 1-4.

[73] S. Heimlicher: Towards Resilience Against Node Failures in Overlay Multicast Schemes: In Communications Systems Research Group, Swiss Federal Institute of Technology Zurich, 11th. August 2004.

[74] Ying Cai and Jianming Zhou: An Overlay Subscription Network for Live Internet TV Broadcast: In IEEE Transactions On Knowledge and Data Engineering, Vol. 18, No. 12, December 2006: IEEE 2006, IEEE Computer Society.

[75] Information Technology—Relayed Multicast Control Protocol (RMCP)—Part 1: Framework: Technologies de l'information—Protocole de multidiffusion relayé (RMCP)—Reference Number ISO/IEC 16512-1:2005(E), ISO/IEC 2005.

[76] C. Casetti, M. Gerla, S. Mascolo, M.Y. Sanadidi, and R. Wang: TCP Westwood: End-to-End Bandwidth Estimation for Enhanced Transport over Wireless Links, Wireless Networks 8, 467–479, 2002, Kluwer Academic Publishers.

[77] Jeonghoon Park: NS Simulator for beginners, Chapter 2. NS Simulator Preliminaries, Networking Laboratory, Sungkyunkwan University, Korea, 2007-04-14.

[78] Federico Montesino-Pouzols: Comparative Analysis of Active Bandwidth Estimation Tools, Instituto de Microelectrnica de Sevilla (IMSE-CNM)—CSIC, Avda. Reina Mercedes s/n. Edif. CICA. E-41012 Seville, Spain

[79] James F. Kurose, Keith W. Ross, Computer Networking: A Top-Down Approach Featuring the Internet, 1st edition, Addison Wesley Longman, Inc., 2001.

[80] William Starlings, Data and Computer Communication, 6th edition, Prentice Hall International, Inc, 2000.

[81] J. R. Freer, Computer Communications & Networks, 2nd edition, UCL Press Limited, 1996.

[82] Network Simulator version 2 Programming Manual.

[83] Brian "Beej" Hall, Beej's Guide to Network Programming Using Internet Sockets, 1995-2001.

[84] Felix von Leitner, Scalable Network Programming: The Quest For A Good Web Server (That Survives Slashdot,), 2003-10-16.

[85] W. Richard Stevens, Unix Network Programming: Networking APIs: Sockets and XTI, 2nd Ed.

[86] Abderrahim Benslimane: Multimedia Multicast on the Internet, British Library Cataloguing-in-Publication Data

[87] Doru Constantinescu: Overlay Multicast Networks: Elements, Architectures and Performance: Publisher: Blekinge Institute of Technology, Printed by Printfabriken, Karlskrona, Sweden 2007, ISBN 978-91-7295-125-9.

[88] John Pearson, Basic communication theory, Prentice Hall, 1992.

[89] F.G. Strenler, Introduction to Communication Systems, 3rd Ed., Addison. Wesley, 1990.

[90] Herbert Taub & Donald L. Schilling, Principles of Communication Systems,2nd Edition, Mc. Graw Hill, 1986.

[91] Bruce Carlson, Communication Systems: An Introduction to Signals and Noise in Electrical Communication", McGraw Hill, 3rd Ed, 1986.

Tutorials, Quizzes and Tests

Appendix A: Tutorials

A.1 Tutorial One

1. Describe the basic elements of a simple communication system and of radio communication system.

2. Discuss some examples of non electric means of information transmission

3. Try to understand the terms: distortion, attenuation, interference, and noise as applicable in telecommunication engineering. Get clear technical meanings of information, messages, and signals in telecommunication engineering.

4. Why is it preferred to convert information into electrical signals before being transmitted.

5. What do signals mean in telecommunications? What is a spectrum? Why are we interested in the frequency-domain description of the signals?

6. Discuss the fundamental tool(s) used to analyze spectra.

7. Discuss, with the help of some common examples, the two main types of signals: analogue and digital.

8. Explain (i) periodic signals (ii) non periodic signals.

A.2 Tutorial Two

1. Explain the following:

(i)   Time-domain and frequency-domain representation of a signal
(ii) Fourier transform pair

2. Find the Fourier transform of the following:
   (i) $e^{-at} u(t)$
   (ii) $e^{-a|t|}$
   (iii) $\delta(t)$
   (iv) $G_p(t) = \begin{cases} 1, & for \quad t \le \tau/2 \\ 0, & for \quad t > \tau/2 \end{cases}$
   (v)   f(t) = A (constant)
   (vi) f(t) = u(t)

   where u(t) is unit step function
   Sketch their time-domain and frequency-domain representations.

3. Find the Fourier transform of the signum function

   $Sign(t) = \begin{cases} 1, & for \quad t > 0 \\ -1, & for \quad t \le 0 \end{cases}$

4. Find the Fourier transform of:
   (i)   Cos $\omega_0 t$
   (ii) Sin $\omega_0 t$
   (iii) $e^{j\omega_0 t}, -\infty < t < \infty$

5. If $f(t) \longleftrightarrow F(\omega)$, then prove that
   (i) $df/dt \longleftrightarrow (j\omega) F(\omega)$

   (ii) $f(at) \longleftrightarrow (1/| a |) F(\omega/a)$

   Discuss the significance of the property (ii) of Fourier transform.

6. State and prove **convolution theorem.** Explain the meaning and the significance of convolution integral. Given $f_1(t)$ and $f_2(t)$, show that convolution is commutative.

7. Evaluate the Fourier transform of a trapezoidal function **f(t)** shown below:

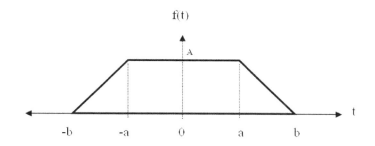

8. Find the Fourier transform of the functions **f(t)** shown in figure (i) and (ii) below:

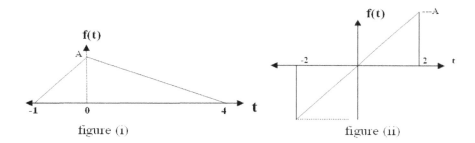

figure (i)   figure (ii)

9. Verify the following using time convolution theorem
   (i) $f(t)*\delta(t-T) = f(t-T)$
   (ii) $f(t-t_1)*\delta(t-t_2) = f(t-t_1-t_2)$
   (iii) $\delta(t-t_1)*\delta(t-t_2) = \delta(t-t_1-t_2)$

10. (a). Explain the popularity of Fourier transform in signal analysis.
    (b). Prove that the convolution in time domain leads to multipli-cation in frequency domain.

A.3 Tutorial Three

1. (a). What is the physical interpretation of power spectral density spectrum?
   (b). Consider a power signal f(t) with a power spectral density of the signal $S_f(\omega)$. Find the power spectral density of the signal df/dt.

2. (a). A carrier is amplitude modulated by the base-band signal $e_m(t)$ to produce

$$S(t) = A_c (1 + k_a e_m(t)) \text{ Cos } 2\pi f_c t$$

Assuming $e_m(t)$ to be an arbitrary time varying analog signal, sketch the resulting signal in the time domain when $k_a e_m(t) < 1$.

3. A carrier $e_c(t) = A_c \text{ Cos } \omega_c t$ is amplitude modulated by a tone sinu-soidal voltage
$e_m(t) = A_m \text{ Cos } \omega_m t$ to produce $S(t) = A_c (1 + \mu \omega_m(t)) \text{ Cos } \omega_c t$. Sketch the waveform $S(t)$ for
(i) $\mu < 1$. (ii) $\mu = 1$, (iii) $\mu > 1$. Indicate the amplitude level of the resulting signal. Express μ in terms of maximum and minimum values of the carrier amplitude.

4. A transmitter radiates 9 KW of power with carrier un-modulated and 10.125 KW when the carrier is modulated. Calculate the depth of modulation. If another sine wave corresponding to 40% modulation is transmitted simultaneously, determine the total power radiated. Derive the formula used.

5. A 10kW radio broadcasting transmitter is modulated by 1 KHz, f_m tone. Calculate the power in carrier and side-bands when the modulation is (i) 10% (ii) 60% (iii) 100%.

6. A broadcast transmitter radiates 10.8 KW of modulated power when the carrier power of 9.6 KW is modulated by a sinusoidal voltage. Calculate the modulation index if another sine wave modulates the same carrier simultaneously to a depth of 30%. Calculate the total radiated power.

7. A wave has a total transmitted power of 5 KW when modulated at 90%. What is the power in SSB wave if it is to have the same power as in the two side-bands.

8. What is the shortcoming of AM? What is DSBSC? Describe (a). Balanced modulator (b). Ring modulator.

9. (a). What is SSB transmission? What are its disadvantages? Mention the method of SSB generation.
   (b). Describe with block diagram the working of any one method of SSB generation.

10. What is the saving in signal power in the case of 75% modulated DSBSC AM wave?

A.4 Tutorial Four

11. Derive an expression for an FM wave when the modulating signal is a single-tone signal.

12. What is modulation index, β, in FM? Write down an expression in terms of frequency deviation and modulating frequency.

13. What is the difference between narrow band and wide band frequency modulations? Draw a block diagram of a method for generating a narrow band FM signal.

14. Suggest a method to convert NBFM to WBFM.

15. Consider an FM system where the modulating signal is a sinusoid of ± 2V peak-to-peak at a frequency of 20KHz. The modulation has a frequency sensitivity of 50KHz/V. Determine the significant bandwidth of the FM signal using CARSON's RULE. Is this NBFM or WBFM? Assume that the carrier frequency is 500KHz.

16. Describe (i) Direct method and (ii) Indirect method of generating FM waves.

17. A certain FM signal is represented by $v(t) = 10\sin(10^8t + 15\sin2000t)$ volts, where t is in seconds. Find the parameters of the FM wave.

18. Compare the FM system with the AM system from the point of view of noise performance and bandwidth saving.

19. What is a superheterodyne receiver? Why is it so called? Discuss the reactive merits over TRF receive.

20. Describe, with a block diagram, the working of FM receiver.

A.5 Tutorial Five

1. Describe the operation of amplitude limiter and automatic frequency control circuits in an FM receiver.

2. Describe how the use of pre-emphasis and de-emphasis circuits in FM system improves the noise performance.

3. Describe the generation of SSB wave by
   (a) Frequency discrimination method
   (b) Phase discrimination method
   (c) Weaver's method.

4. What are the physical significance of mean and variance of a random variable? Use these concepts to analyze the random noise in telecommunications.

5. A random process, noise, $n(t)$ has a power spectral density $\sigma(f) = \eta/2$ for $-\infty \le f \le \infty$. The random process is passed through a low pass filter which has a transfer function $H(f) = 2$ for $-f_M \le f \le f_M$ and $H(f) = 0$ otherwise. Find the power spectral density of the waveform at the output of the filter.

6. If the noise power in a channel is 0.1dBm and the signal power is 20mW, what is the signal to noise ratio?

7. Explain noise and classify sources of noise broadly. Discuss the idea of thermal noise. Show that the spectral density of thermal noise is constant at $G_v(f) = 2RkT$ (Volts)^2/Hz.

8. Express the spectral density of white noise, $G(f)$, in terms of positive-frequency power density, η. What are the factors that this η

depends? Deduce the white noise auto-correlation, $R(\tau)$, in terms of η and $\delta(\tau)$.

9. Define the noise temperature, T_N, of any white noise source; thermal or non thermal. What is the physical significance of T_N?

10. Show that in AM system $4S_0 = S_i$ independently of whether a single or many spectral components are involved. Deduce SNR at the output of an AM system. What does SNR tells about the performance of a communication system? Show that SNR for SSB-SC is exactly as for DSB-SC.

11. Express S_0 and N_0 and hence compute the output SNR of FM system. Repeat the computation of SNR by now considering that the modulating signal, say m(t), is sinusoidal and produces a frequency deviation Δf.

12. Compare angle and linear modulation in terms of the figure of merit, γ. What are the assumptions to consider in ideal comparison of the said two modulations? Show that each increase in bandwidth by a factor of 2 increases $(\gamma_{FM})/(\gamma_{AM})$ by 6dB.

13. Explain noise figure, F, and the assumptions to be made in analyzing the noise figure of the two port equivalent circuits.

14. Explain the noise figure, F, and the equivalent temperature, T_e, for the case of a cascade of k stages.

15. State the sampling theorem. Prove with appropriate derivation that a function f(t) can be recovered from its samples.

16. What is multiplexing? Draw the block diagram of frequency division multiplexing system and explain it clearly.

17. Estimate the bandwidth requirements of a single satellite that is to support 10 lakh telephone conversations simultaneously?

18. Verify the sampling theorem for a band limited signal and for a band pass signal.

19. What is TDM? What are the merits and demerits of TDM? Make comparison with FDM. What is the major practical problem of FDM? What is the primary cause? What are the advantages of TDM?

20. Explain with a practical AT &T application system the notion the "Excessive guard band-band width can be avoided by grouping".

21. Why is synchronization a critical consideration in TDM? Explain Brute-force type of synchronization.

22. Suppose 5 data channels are needed with the following minimum sampling rates:
$x_1(t)$: 3000Hz; $x_2(t)$: 700Hz; $x_3(t)$: 500Hz; $x_4(t)$: 300Hz; $x_5(t)$: 200Hz. Use an efficient scheme of multiplexing these 5 channels and show that the total output signaling rate is 6kHz.

23. What must be the bandwidth of the communication channel that will allow N signals to transmit simultaneously using PAM-TDM?

24. What is pulse modulation? Explain its advantage over CW modulation. Enumerate the types of pulse modulation.

25. Discuss in detail the (i) Pulse Amplitude Modulation (PAM) (ii) Pulse Duration? Width Modulation (PD/WM) and (iii) Pulse Position Modulation (PPM).

26. The signal $g(t) = 10 \, Cos(2\pi t) \, Cos(200\pi t)$ is sampled at the rate of 250 samples per second.
    (a) Determine the spectrum of the resulting sampled signal
    (b) Specify the cutoff frequency of the ideal reconstruction filter so as to recover g(t) from its sampled version.
    (c) By treating g(t) as a band-pass signal, determine the lowest permissible sampling rate for the signal.
    (d) What is the Nyquist rate for g(t)?

27. A band pass signal has a spectral range that extends from 20 to 82 kHz. Find the acceptable range of the sampling frequency (f_s).

28. The TDM system is used to multiplex the four signals $m_1(t) = \text{Cos}(\omega_o t)$; $m_2(t) = 0.5\text{Cos}(\omega_o t)$; $m_3(t) = 2\text{Cos}(2\omega_o t)$ and $m_4(t) = \text{Cos}(4\omega_o t)$.
    (a) If each signal is sampled at the same sampling rate, calculate the minimum sampling rate, f_s.
    (b) What is the commutator speed in r.p.m?

29. Three signals; m_1, m_2 and m_3 are to be multiplexed. The signals m_1 and m_2 have a 5kHz bandwidth each and m_3 has a 10kHz bandwidth. Design a commutator-decommutator switching system so that each signal is sampled at its Nyquist rate.

30. Describe a block diagram of the circuit(s) required to form a pulse duration modulated signal and describe their operation

31. Draw the block diagram of the circuit required for forming a pulse position modulated signal and describe the principle of operation.

32. Describe, with suitable example, how sampling, quantization and encoding of a signal are performed.

33. Define quantization noise and quantization error. Derive the expression for the mean square quantization in terms of the step size of quantization, S.

34. Explain, with suitable block diagram, a PCM. How are quantization and coding done?

35. Calculate the output signal-to-quantization noise ratio in dB for a PCM and give your comments on the result.

36. Describe different methods of multiplexing the PCM signals.

37. What are the difficulties in PCM systems? How can they be over-come by DPCM? Support your answer with the help of a typical example.

38. Describe companding and companding curves for PCM. PCM is marvelous, but other systems are still in good use. Explain.

39. Analyze the Bell T1 digital transmission system with 24 PCM channels used in North America. How many bits per second will be sent?

40. A binary channel with r_b =36,000bps is available for PCM voice transmission. Find appropriate values of v, q, and f_s assuming that $W \cong 3.2kHz$. All the notations are in standard uses.

41. What is data transmission? Describe, with a simple analog base-band transmission system, the baseband transmission. Relate the signals S_x, S_T, S_R, S_D, g_T, g_R, and L, where all the notations are in standard uses.

42. Derive $(S/N)_D$ and show that the receiver gain g_R amplifies signal and noise equally.

43. Show that by dividing the path/link into m equal sections, each having loss L_1, there is a potential improvement of $(S/N)_D$ by a fac-tor of $L/(mL_1)$

44. Express $(S/N)_{D, dB}$ in terms of S_R, T_N, and T_o. What is the nominal message bandwidth, W?

45. Let L for a cable system is 10^{14} and that T_N is five times T_0. Let that the transmission is high-fidelity audio transmission such that W = 20kHz. Let $(S/N)_D$ demanded is greater than or equal to 60dB. Show that the demanded transmitted power is not bearable on a signal transmission cable without inserting a repeater. What will be L_1 in terms of dB after inserting a repeater at the midpoint. What will be the resulting improvement factor? What will be the new transmitted power requirement? Is the new power a realistic value?

46. Explain the main causes of errors in the data stream during transmission? Clarify the terms; error detection, error correction, and error control.

47. Explain parity-check codes, constant ratio codes, and redundant codes.

48. Detecting errors in telecommunications data transmission is of little use. Explain.

49. Explain the common retransmission techniques. Give details descriptions on "ACK" and on "NAK".

50. Explain Forward-Error-Correcting Codes (FEC).

A.6 Tutorial Six

1. Compare digital technology with analog technology. Where does the digital technology become of particular importance? Why are military and commercial telecommunications systems going digital? Explain the cost and the disadvantage of digital communication systems.

2. Describe the idea of modern digital communications system stressing on the 5 essential elements: Format, Pulse Modulation, Band-pass Modulation, Synchronization, and Detection blocks.

3. Explain the digital versus analog performance criteria. A figure of merit for digital communication system can lead to sometimes 100% error–free performance and sometimes 100% transmission/reception error. Explain.

4. PAM, PTM, and PCM are all the forms of Pulse Modulation. Categorize them and make clear with those that have connection with AM/FM.

5. Given the source information, underline the procedures to produce the corresponding PCM code words. Mention the relationship be-

tween the performance (sensitivity, selectivity, and fidelity) and the number of quantization levels.

6. Why is PCM appears in both, formatting and in baseband signaling blocks? Differentiate between PCM and PCM waveform.

7. PCM waveforms are classified into four major groups: NRZ, RZ, Phase Encoded, and Multilevel Binary. Describe each group and the subgroups with one, and the same, example.

8. From (7) above, explain the 6 parameters worth to examine in choosing a PCM waveform.

9. Describe the generation of PCM insisting on the need of an L-level quantizer and on parallel-to-serial converter.

10. Describe the PCM demodulation and explain the statement "Perfect message reconstruction is impossible in PCM, even when random noise has no effect".

11. DPCM was invented to overcome the difficulties in PCM. Explain, how? Give a typical example to support your explanation.

12. Explain Delta Modulation (DM) of base-band signal with the help of block diagram. Describe the slope over-load error in connection to DM.

13. Write down the mathematical relation to be achieved between the sampling frequency (fs_s) and the frequency (f) of a sinusoid having amplitude (A) to avoid slope over-load in DM system.

14. What are the limitations of DM system? Suggest and explain arrangement(s) to overcome the limitations.

15. Describe the generalized idea of Adaptive Delta Modulation (ADM). Make clear the expression to select the variable step sizes, S(k) at sampling time k.

16. Write the general analytic expression for PSK. Explain the case of BPSK. Why can BPSK be thought of as an AM system? Explain the nominally fixed phase shift, θ, and explain the factors influencing θ.

17. Describe the BPSK and the DPSK schemes of modulation. Describe how BPSK signals can be generated and recovered. Explain the two main types of interference in BPSK.

18. The data b(t) consists of the bit stream 001010011010. Assume that the bit rate f_b is equal to the carrier frequency f_0 and sketch $v_{BPSK}(t)$

19. Explain clearly the DPSK scheme. Also describe how DPSK signals can be generated and recovered. How is DPSK superior to PSK system?

20. The bit stream d(t) is to be transmitted using DPSK. If d(t) is 001010011010, determine b(t).

A.7 Tutorial Seven

Given an analog music passage signal, with excursion limited to 0 to 4V and the ten samples taken each after every time period, T and the analog equivalent levels given in volts as: 0.15, 0.45, 0.65, 1.80, 2.15, 3.55, 1.97, 1.20, 0.56, and 0.22.

(i)   Obtain, showing all the necessary procedure, the ten code words of the samples employing sixteen levels of quantization.

(ii) If the channel transmission data rate is 10Mbps, how many microseconds will be enough to transmit the given passage assuming that each sample is embedded with one start and one stop bit.

(iii) Produce the whole data string that is going to be transmitted taking into consideration that, to simplify the demodulation process, the gross code words are going to be sent using binary-back-to-front fashion.

(iv) What will be the efficiency and the percentage overhead error of this transmission?

(v) Now assume that each passage of five samples makes frame and each frame is accompanied by one and a half start bit and one stop bit. What will be the new percentage overhead error and what will be the percentage time saving?

(vi) Try to make a theoretical analysis for the case when we have 256 quantization levels and all other parameters (ii, iii, iv, v) being the same wherever applicable.

A.8 Tutorial Eight

1. Differentiate among ASK, FSK, and PSK. What do you understand by QPSK? How is it used to increase the data rate in a communication link?

2. What do you understand by bandwidth efficiency? Compare the same for ASK, FSK, PSK, and QPSK.

3. Explain QPSK system stressing on the basic principles of the transmitter and the receiver sides. Describe the function of a detector for QPSK system.

4. Describe the mechanism used by DQPSK signal generating system.

5. Explain the FSK scheme. Describe with proper block diagram how FSK signals can be generated and recovered. Give a comparison between PSK and FSK.

6. Explain the M-ary PSK (MPSK) system and the representation of the waveforms used to identify the symbols. Justify the fact that "Increasing the number of bits, N, per symbol leads to progressively narrowing of the bandwidth."

7. Describe abstract schemes of superficial MPSK transmitter and receiver.

8. Explain the general expression for the FSK modulation. Give the conceptual representation of the FSK scheme.

9. Explain the BFSK signals if the binary data waveform is given as d(t) and the amplitude is given in terms of power, P_s.

10. How to generate the BFSK signal? Make use of the new derived bipolar variables.

11. Describe the BFSK signal receiver system and explain the uses of a bit synchronizer and an integrator during sampling stage.

12. Compare BFSK and BPSK starting from the equation representing BFSK.

13. Explain, with block diagrams, the M-ary FSK transmitter and receiver. Explain the differences between the MFSK and MPSK in terms of bandwidth and P_E.

14. Explain the functional blocks of a binary QAM. What is relation between the transmission bandwidth and the bit rate for the case of QAM?

15. Why using QASP while BPSK, QPSK, and MPSK can be used for transmission, too?

16. Describe the QASK signal generation. Compare the bandwidth of QASK from that of MPSK.

17. Describe the QASK receiver and show that a signal at frequency $4f_0$ will be recovered even if A_e and A_o are not of fixed values.

18. What does multiplexing mean? Where is it used? Compare between the FDM and the TDM. What do guard bands mean? Why are they required?

19. Differentiate among Group, Supergroup, Mastergroup and Jumbo-group in AT & T standards.

OMAR FAKIH HAMAD

20. Briefly explain Bell system T1 carrier system.

21. Following 11 sources are to be multiplexed:
    Source 1: analogue 2 kHz bandwidth
    Source 2: analogue 4 kHz bandwidth
    Source 3: analogue 2 kHz bandwidth
    Source 4 to 11: digital 7200 bps synchronous.

    Implement a scheme to generate a TDM signal of 128 kbps.

22. Following 12 sources are to be multiplexed.
    Source 1: analogue 1 kHz bandwidth
    Source 2: analogue 1 kHz bandwidth
    Source 3: analogue 2 kHz bandwidth
    Source 4: analogue 4 kHz bandwidth
    Source 4 to 12: digital 7200 bps synchronous.

    Implement a scheme to generate a TDM signal of 128 kbps.

23. What do you mean by average information or entropy and information rate? Give the expressions for them.

24. State and explain Shannon's theorem regarding channel capacity.

25. State Shannon-Hartley theorem regarding the capacity of Gaussian channel.

26. Message Q_1, Q_2, ..., Q_M have probabilities P_1, P_2, ..., P_M of occurring.
    (a) Write an expression for H
    (b) If $M = 3$, write H in terms of P_1, and P_2, using the result that $P_1 + P_2 + P_3 = 1$.
    (c) Find P_1 and P_2 for H_{max}.

27. From 26 above, show that the range of H is limited to 0 and $\log_2 M$, i. e., $0 \leq H \leq \log_2 M$.

28. A Gaussian channel has a 1 MHZ bandwidth. If the Signal-power-to noise power spectral density, $S/N = 10^5$ Hz, calculate the channel capacity, C and the maximum information transfer rate, R.

29. Plot channel capacity, C versus B, with constant S/N for the Gaussian channel.

30. You are given six messages m_1, m_2, m_3, m_4, m_5, and m_6 with probabilities $P(m_1) = 1/3$, $P(m_2) = 1/8$, $P(m_3) = 1/12$, $P(m_4) = 1/4$, $P(m_5) = 1/12$, and $P(m_6) = 1/8$. Find the Fano's minimum redundancy code for this set of messages and the efficiency.

31. Solve problem 29 using the technique of Huffman coding. Compare the efficiency between the Shannon-Fano's and the Huffman's coding techniques.

32. Give technical short notes on: source coding, channel coding, block codes, convolutional codes hamming distance, and BCH code.

33. For a source emitting symbols in independent sequences, show that the source entropy is maximum when the symbols occur with equal probabilities.

34. A black and white TV picture consists of 525 lines of picture information. Assume that each line consists of 525 picture elements and that each element can have 256 brightness levels. Pictures are repeated at the rate of 30/sec. Calculate the average rate of information conveyed by a TV set to a viewer.

35. The output of an information source consists of 128 symbols, 16 of which occur with a probability of 1/32 each and the remaining 112 occur with a probability of 1/224 each. The source emits 1000 symbols/sec. Assuming that the symbols are chosen independently, find the average information rate of this source.

36. Comment on why an error detection scheme is desirable in a data communication system. Describe the parity check method of error detection. What are its limitations?

37. What is meant by data rate or channel capacity of a communication link? What are the factors affecting it?

38. What does hamming distance mean? What is its significance in error detection schemes?

39. Estimate the rate of transmission for a channel used to connect 20,000 telephone subscribers (use voice bandwidth, BW = 4kHz and 1 sample = 8 bits).

40. Give explanation on the probability of error, P_E, in digital tele-communications systems.

Appendix B: Quizzes

B.1.1 Quiz One

1. (a). Describe, with examples, the terms: analog signal and digital signal.
   (b). Explain analog transmission and digital transmission.

2. (a). Explain briefly the terms: distortion, attenuation, interference, and noise.
   (b). Give clear technical meanings of information, messages, and signals.

3. (a). Give a very compact definition of the term telecommunications.
   (b). Describe a very simple telecommunications system.

4. Why do telecommunications engineers give special attention to spectra?

5. Explain the following:
   (a). Time-domain and frequency-domain representation of signals.
   (b). Fourier-transform pair.

B.1.2 Make-up Quiz One

1. (a). Describe, with examples, the terms: periodic signal and aperiodic signal.
    (b). Point out the main concerns of modern telecommunications systems.

2. (a). Explain briefly the terms: distortion, attenuation, interference, and noise.
    (b). Of great importance, in telecommunications, is the amount of information the message contains. Clarify the terms; signal, information, and message.

3. Describe a simple communication system with input and output transducers mentioning the three essential elements of any communication system.

4. Why frequency-domain representation of signals is more preferred than time domain-representation?

5. (a). Express $f(t)$ by the Fourier Series. Remember to write the equations for the three coefficients.
    (b) Explain Fourier-transform pair.

B.2.1 Quiz Two

1. Find the frequency—domain representation of the square—wave signal that varies between +**A** and -**A** with the pulse duration, $\tau = T/2$, where T is the periodic time as shown below. Mention any assumption you wish to make to facilitate the computation.

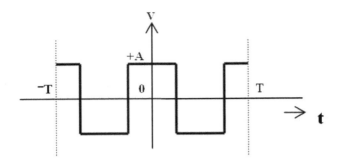

2. Find the Fourier transform of the following common functions:
   (i) $a^n u(t)$ (iii) $\delta(t)$

3. Describe the frequency domain representation of a single rectangular pulse, f(t), with amplitude A and duration τ centered at t = 0. Say something relating the spectral "width" with the τ.

4. The pulse in (3) above is delayed by T_0. Express the resultant $F(\omega)$. Explain the theorem you have applied.

5. Under what condition(s) does a periodic signal can be considered as a non-periodic signal?

B.2.2 Make-up Quiz Two

1. Find the Fourier components of the following typical waveform used in practice:

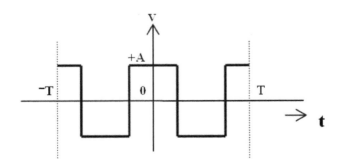

2. Find the Fourier transforms of the following signals for an arbitrary constant, t'.
   (i) $a^{(t-t')}u(t-t')$ (ii) $\delta(t-t')$

3. Describe the frequency domain representation of the single pulse $v(t) = A\pi(t/\tau)$.

4. Explain the shift theorem of transform. Using the results of (3) above, determine the Fourier transform of the shifted pulse below:

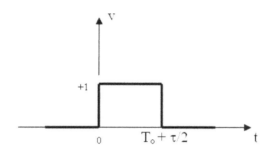

5. Explain the convolution theorem and its significance in time and frequency domain relationships of the signals.

B.3.1 Quiz Three

1. Explain the superiority of FM over AM. What is DSB-SC? Describe either Balanced modulator or Ring modulator.

2. Describe the need of modulation. Write a short technical description on envelope detector in an AM indicating all the significant voltage levels.

3. Describe, with block diagram, coherent or synchronous detection of DSB-SC?

4. The equation of an FM voltage is e = $6Sin(10^6t + 45Sin10^3t)$. Calculate:
   (iii) carrier frequency   (ii) modulating frequency

| (iv) modulation index | (iv) deviation and power dissipated in a 50 ohm resistor. |

B.3.2 Make-up Quiz Three

1. Explain the five technical benefits and applications of modulation. Explain the two waveforms involved during modulation. Describe Cowan modulator.

2. Give a short technical description on Independent Side Band (ISB) transmission system, where each side band carries a different message.

3. An AM transmitter has an anode-modulated class-C output stage in which an AF sine wave of 3kV peak value is developed across the secondary of the modulating transformer in series with the 5kV HT supply. The stage has an anode efficiency of 75% and delivers 1.5kW of carrier power into the tank circuit. Calculate:
(a) the modulation index
(b) the mean anode current
(c) the power supplied by the modulator
(d) the total RF power delivered to the tank circuit.

4. A certain FM signal is represented by $v(t) = 10Sin(10^8t + 15Sin2000t)$ volts, where t is in seconds. Find the parameters of the FM wave.

B.4 Quiz Four

1. Explain QPSK system stressing on the basic principles of the transmitter and the receiver sides. Describe the function of a detector for QPSK system.

2. Describe the mechanism used by DQPSK signal generating system.

3. Explain the M-ary PSK (MPSK) system and the representation of the waveforms used to identify the symbols. Justify the fact that

"Increasing the number of bits, N, per symbol leads to progressively narrowing of the bandwidth."

4. (a) Explain the BFSK signals if the binary data waveform is given as d(t) and the amplitude is given in terms of power, P_s.
   (b) How to generate the BFSK signal? Make use of the new derived bipolar variables.

5. (a) Why using QASK while BPSK, QPSK, and MPSK can be used for transmission, too?
   (b) Following 11 sources are to be multiplexed:

   >   Source 1: analogue 2 kHz bandwidth
   >   Source 2: analogue 4 kHz bandwidth
   >   Source 3: analogue 2 kHz bandwidth
   >   Source 4 to 11: digital 7200 bps synchronous.

   Implement a scheme to generate a TDM signal of 128 kbps.

Appendix C: Tests

C.1.1 Test One

1. Give the definition and the significance of each of the following terms as used in telecommunications: distortion, attenuation, noise, interference information, message, signal, spectrum

2. Describe, with example, a simple input-transducer-output-transducer communication system mentioning the three essential elements of any communication system.

3. Explain (a). Time-domain and frequency-domain representation of signals.
   (b). Fourier-transform pair.

4. Verify the following using convolution theorem
   (i) $f(t)*\delta(t-T) = f(t-T)$
   (ii) $f(t-t_1)*\delta(t-t_2) = f(t-t_1-t_2)$
   (iii) $\delta(t-t_1)*\delta(t-t_2) = \delta(t-t_1-t_2)$

5. Use Fourier series to express the periodic signal below into sinusoidal components. Show clearly all the necessary steps for a rectangular square wave signal that varies between **+10** and **-10** with the pulse duration, $\tau = T/2$, where T is the periodic time as shown:

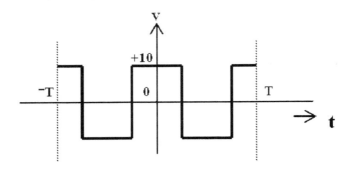

6. Evaluate the Inverse Fourier transform, $h_{lp}(t)$, of an ideal LPF whose impulse frequency response for a time period 2π is given as below:

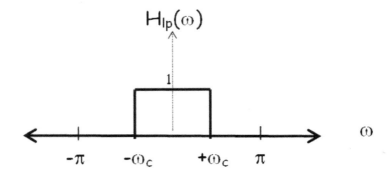

Decide, by justification, whether an ideal LPF is causal or not

7. With all the notations in standard uses in AM, two signals are given as: $v_c=V_cSin\omega_ct$ and $v_m=V_mSin\omega_mt$. Deduce v_c constituted by the three components and having relation with the depth of modulation.

8. Explain a Ring Modulator. Why is it called so? How to produce an SSBSC signal from DSBSC?

9. Explain VSB system.

10. Stressing on Superheterodyne Principle, describe a receiving system for an AM signal.

C.1.2 Make-up Test One

1. Explain two main concerns of a telecommunications engineer. Why converting information into a signal?

2. Describe a typical communication system mentioning the three essential elements. What are the seven main concerns of a modern telecommunications system.

3. Describe (a). Analogue signals (b) Discrete-time signals (c) Digital signal (d) Causal system

4. Explain the representation of a periodic signal by Fourier series.

5. Analyze a rectangular pulse $v(t) = A\pi(t/\tau)$ using Fourier transform in terms of $2\pi f$.

6. Evaluate the Inverse Fourier transform, $h_{lp}(t)$, of an ideal LPF whose impulse frequency response for a time period 2π is given as below:

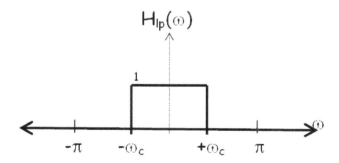

Decide, by justification, whether an ideal LPF is causal or not.

7. Describe the basic principle of AM with two signals given as: $v_c = V_c Sin\omega_c t$ and $v_m = V_m Sin\omega_m t$. Explain the power distribution for the carrier and the side-bands

7. Explain a Balanced Modulator and a Ring Modulator. How to produce an SSBSC signal from DSBSC?

9. Describe the AM spectrum.

10. Describe Superheterodyne Principle as applicable to AM receivers.

C.2.1 Test Two

1. (a) Explain, with at least five reasons, why are telecommunications systems digitized? What is the use of constructing unique codes in digital systems? Explain about the cost and disadvantage(s) of digital communication systems.
   (b) Describe an "ideal modern digital telecommunications system" classifying the essential and optional operational blocks. Explain the need for frequency spreading and the need for multiplexing/multiple access procedures.

2. (a) Given a real time analog voice signal, with excursion limited to the range -2 to +2V, describe the procedure to be followed to obtain seven code words employing eight levels of quantization. The samples are taken after every time period, T and the seven analog equivalent levels are noted in volts as: 0.65, 1.80, 1.15, 0.35, -0.35, -1.2, and -1.7.
   (b) What will be the consequences of decreasing the number of levels from 8 to 4 in the following contexts: fidelity, noise, delay, transmission bandwidth, and bit-pulse-width?

3. (a) Differentiate between PCM and PCM waveform. Make a tree to classify the common known PCM waveforms. Describe, with example(s), the Manchester coding of PCM waveform.
   (b) Why and which code can be of advantageous, each, when the parameter under consideration is: self-clocking, error-

detection, bandwidth compression. What are the other, important parameters to be examined?

4. (a) Describe the generation of PCM and the reception of PCM. Even if there is no random noise effect, perfect message reconstruction is impossible in PCM. Justify.
    (b) What are the limitations of PCM over DPCM? Draw the schematic block diagram to show the basic principle of DPCM. The receiver and the transmitter parts need to be shown.

5. (a) Explain, with examples, the idea of slope-overload error. How does the Adaptive Delta Modulation (ADM) reduce the effects caused by fixed step size?
    (b) Describe the generation of DPSK. Given an in input d(t) = 00100110011110, draw the equivalent logic waveforms of d(t), b(t -1), and b(t) for the case of DPSK.

C.2.2 Make-up Test Two

1. (a) Give a brief explanation on each of the following digital communication nomenclature: information source, textual message, character, binary digit, bit stream, symbol, and digital waveform. Explain the criteria for digital versus analog performance.
    (b) Describe an "ideal modern digital telecommunications system" classifying the essential and optional operational blocks. Explain the need for frequency spreading and the need for multiplexing/multiple access procedures.

2. (a) Given an analog voice signal, with excursion limited to -2 to +2V, obtain the code words of the samples employing sixteen levels of quantization. The samples are taken after every time period, T and the analog equivalent levels to be encoded are given in volts as: 0.65, 1.80, 1.15, 0.35, -0.35, -1.2, and -1.7.
    (b) If each code word is accompanied by a binary bit "1" as a start bit and a binary bit "0" as a stop bit, What string will be transmitted after 7T? Assume that the code words are being sent as a binary number back to front to simplify the demodulation procedure.

3. (a) Describe the three classes of RZ type of PCM waveforms and the Manchester coding of PCM waveform. Explain Duo-binary signaling of PCM waveforms,
   (b) Explain, with an example, the basic principle of linear Delta Modulation (DM). Why does the linear DM find almost no applications in real systems? Give an example.

4. (a) Give the general analytic expression for PSK explaining all the significant terms constituting it. Justify the idea that "BPSK can be thought of as an AM signal.
   (b) Draw the schematic block diagram to generate the carrier at the demodulator and recover the BPSK baseband signal. What are the merits of DPSK and DEPSK over BPSK?

5. (a) Explain the recovering of the data bit stream from the DPSK. Mention the advantage(s) and the disadvantage(s) of the DPSK over the BPSK
   (b) Given an input d(t) = 00100110011110, show that the equivalent logic waveforms for d(t), b(t -1), and b(t) for the case of DPSK systems are consistent with one another.

=====